Mattia Del Giacco

Investigation of Fretting Wear of Cladding Materials in Liquid Lead

Karlsruher Forschungsberichte aus dem
Institut für Hochleistungsimpuls- und Mikrowellentechnik

Herausgeber: Prof. Dr.-Ing. John Jelonnek

Band 3

Investigation of Fretting Wear of Cladding Materials in Liquid Lead

by
Mattia Del Giacco

KIT Scientific Publishing

Dissertation, Karlsruher Institut für Technologie (KIT)
Fakultät für Maschinenbau, 2012
Referenten: Prof. Dr. rer. nat. Oliver Kraft, Prof. Dr.-Ing. John Jelonnek

Impressum

Karlsruher Institut für Technologie (KIT)
KIT Scientific Publishing
Straße am Forum 2
D-76131 Karlsruhe
www.ksp.kit.edu

KIT – Universität des Landes Baden-Württemberg und
nationales Forschungszentrum in der Helmholtz-Gemeinschaft

KIT Scientific Publishing 2013
Print on Demand

ISSN 2192-2764
ISBN 978-3-86644-960-2

Foreword of the Editor

Compatibility of structural materials (steels) with liquid-lead (Pb), foreseen as a promising coolant for use in future Generation IV reactors and so-called "Transmutors", is under strong investigation within the frame of international projects since long time. Dissolution attack and severe oxidation are the corrosion mechanisms that affect structural materials in contact with liquid-lead. Surface modification by aluminum alloying using the GESA-process is one promising option to counteract compatibility problems of steels that are supposed to be used at high temperatures, e.g. fuel claddings in liquid-lead. So far, experimental results regarding fretting at fuel claddings and heat exchanger tubes by flow induced vibrations are lacking. Beside basic tribological phenomena additional stress arise from the specific operating conditions in nuclear reactors and the coolant environment. In particular, the interaction between the fretting wear and the corrosion has to be considered.

Based on a comprehensive study, Dr.-Ing. Mattia DelGiacco designed and constructed a fully instrumented fretting test facility. Using this facility fretting tests in liquid-lead has been performed for the first time at relevant operating conditions regarding temperature and oxygen content of the Pb. Extensive tests using fuel clad materials (T91, 15-15Ti and GESA) are performed varying the most important parameters like amplitude, load, number of cycles and temperature. The effect of the individual parameters on the fretting behavior and the interactive influence of fretting and corrosion in Pb are shown. E. g. the steel 15-15 Ti containing Nickel suffers from dissolution attack under fretting conditions at significant lower temperatures than without additional stress. Long-term tests lasting 900 h and including $3x10^7$ cycles are used to determine the long term behavior and the specific wear coefficient and the wear rate.

With this PhD thesis Dr.-Ing. Mattia Del Giacco demonstrates a first of its kind test apparatus that allows to perform fretting tests in liquid lead with high accuracy and reproducibility also for long-term tests. He investigated for the first time the fretting behavior of the relevant fuel clad materials in liquid-lead. By evaluating the specific wear coefficient and adding the concept of fretting maps he was able to predict the tolerable operating conditions for fretting wear.

Investigation of Fretting Wear of Cladding Materials in Liquid Lead

Zur Erlangung des akademischen Grades

Doktor der Ingenieurwissenschaften

der Fakultät für Maschinenbau

Karlsruher Institut für Technologie (KIT)

vorgelegte

Dissertation

von

Mattia Del Giacco

aus Carate Brianza, Italien

Tag der mündlichen Prüfung:	29. Oktober 2012
Hauptreferent:	Prof. Dr. rer. nat. Oliver Kraft
Korreferent:	Prof. Dr.-Ing. John Jelonnek

Dubium sapientiae initium

(Descartes)

Dedicato a tutte le persone che amo;
con un pensiere speciale rivolto a
mia Nonna Mariuccia scomparsa il 25.05.2010.

Zusammenfassung

Die Entwicklung von Strukturmaterialien, die den Anforderungen während des Betriebs und in Notfallsituationen genügen, ist einer der Hauptaspekte zur Realisierung von bleigekühlten Hochleistungskernreaktoren (zB. LFRs and ADS).

Insbesondere Oxidation und Lösungsangriff sind die Korrosionsmechanismen, die in flüssigem Blei bzw. flüssigen Bleilegierungen auf die Strukturmaterialen einwirken.

Erhebliche Anstrengungen wurden gemacht, um die Beständigkeit der Materialien gegen die genannten Korrosionsmechanismen zu untersuchen und zu verbessern. Eine adequate Korrosionsbarriere wird durch die Bildung einer langsam wachsenden Oxidschicht, die den Korrosionsschutz und einen genügenden Wärmeaustausch (wichtig für Bauteile im Reaktorkern) sichert, erreicht.

Strukturmaterialien und Korrosionsbarrieren müssen hinsichtlich ihrer mechanischen Eigenschaften und ihres Abriebwiderstandes getestet werden. Ein spezifisches Problem von Brennelementhüllrohren und Rohren von Wärmeübertragern ist das Fretting. Diese spezielle Art von Reibverschleiß, die von strömungsinduzierten Schwingungen erzeugt wird, kann die schützende Oxidschicht schwächen und schädigen. Dadurch werden die Korrosionsmechanismen (z.B. der Lösungsangriff) gefördert und die Lebensdauer der Bauteile wird reduziert. Deshalb sind die Untersuchung, Bewertung und die Vorhersage von Frettingsverschleiß wichtig für das Reaktordesign, insbesondere unter Berücksichtigung von Sicherheitsaspekten und ökonomischer Auslegung.

Zu diesem Zweck wurde die FRETHME (FREtting corrosion Test in Heavy liquid Metals) Anlage entworfen und gebaut, die Frettingtests unter typischen Betriebsbedingungen eines bleigekühlten Kernreaktors erlaubt. Die für den Reibverschleiß notwendige Relativbewegung zweier Körper wird durch eine gegen einen starr fixierten Probenkörper bewegte Probe realisiert.

Die wichtigsten Parameter wie Amplitude, Frequenz, angelegte Last und Reibkraft werden während des Frettingtests kontinuierlich beobachtet und aufgezeichnet. Ebenso werden Temperatur und Sauerstoffgehalt des flüssigen Bleis in einem den Betriebsbedingungen eines Reaktors entsprechenden Bereich mit einem dafür konzipierten Heizungsystem und einer Gasversorgungsanlage kontrolliert.

Die FRETHME Anlage wurde verwendet, um erstmals eine Versuchskampagne von Frettingtests in flüssigem Blei durchzuführen. Die Ziele der Versuchskampagne waren, den Verschleißmechanismus zu bestätigen, die Auswirkungen der wichtigsten Parameter auf den Frettingprozess zu untersuchen und die Betriebsbedingungen, bei denen Fretting minimiert wird, zu definieren. Die gegenseitige Wechselwirkung von Fretting und Korrosionsmechanismen (z.B. Tribokorrosion) und der Einfluss des flüssigen Bleis auf den Frettingprozess waren ebenso Gegenstand der Untersuchungen.

Die Materialien für die Versuchskampagne wurden aus den für bleigekühlte Hochleistungskernreaktoren vorgesehenen ausgewählt. Diese Stähle sind: der austenitische 15-15Ti Stahl, der ferritisch-martensitische T91 Stahl und der oberflächenaluminisierte T91 Stahl (GESA-T91).

Vor den Frettingtests in flüssigem Blei wurden einige Experimente in Luft durchgeführt. Die Ergebnisse dieser Tests zeigten eine allgemeine Übereinstimmung mit vorhandenen Datensätzen und demonstrierten die Zuverlässigkeit und die Eignung der FRETHME für die geplanten Untersuchungen.

Die Resultate der Versuche deuten an, dass die Frettingabnutzung hauptsächlich mit der Schwingungszahl, mit der Temperatur (bis zu 550°C) und mit der Amplitude ansteigt. Die Auswirkungen der aufgelegten Last sind korreliert mit der Amplitude der Gleitbewegung. Die Frettingabnutzung

steigt mit steigender aufgelegter Last bis zu einem Extrempunkt an und nimmt dann wieder ab. Je höher die Amplitude der Gleitbewegung ist, desto höher wird der Wert der aufgelegten Last am Wendepunkt.
Sowohl die aufgelegte Last als auch die Amplitude der Gleitbewegung beeinflussen signifikant die Schwere des Frettingverschleißes, weil beide eine entscheidende Rolle bei der Festlegung des Frettingsregimes spielen. Das Wechseln von einem Frettingregime zu einem anderen wird als Erklärung für die nicht-monotone Last- Frettingabnutzung Relation und die nicht-lineare Steigerung der Frettingabnutzung mit der Amplitude vorgeschlagen.
Bezüglich einer möglichen Wechselwirkung Fretting / Korrosion wurde festgestellt, dass Korrosionsmechanismen (Oxidation und Lösungsangriff), denen Stähle in flüssigem Blei ausgesetzt sind, sich verstärkend auf den Frettingsverschleiß auswirken. Im Vergleich zu Experimenten in Luft jedoch ist der Frettingsverschleiß in Blei wegen der Schmierwirkung des flüssigen Bleis reduziert. Außerdem kann die Frettingbewegung die Korrosionsbarrieren (z.B. Oxidschicht) abtragen und damit den Lösungsangriff verstärken.
Unter Berücksichtigung all dieser Aspekte zeigten die GESA-T91 Stahlproben, insbesondere wegen des ausgezeichneten Korrosions- und Abnutzungswiderstandes, das beste Verhalten der getesteten Materialien.
Im Gegensatz dazu litt der 15-15Ti Stahl (wegen des hohen Nickel Gehalts) bei fast allen Testbedingungen (insbesondere bei höheren Temperaturen) unter Lösungsangriff. Obwohl der T91 Stahl generell gegen Fretting beständiger war als der 15-15Ti Stahl, wurde bei einer Temperatur > 500°C die Frettingabnutzung durch die zunehmende Oxidation deutlich verstärkt. Voroxidation der Proben und Anreicherung des flüssigen Bleis mit Ni, reduzieren bei allen Materialien den Verschleiß durch das Fretting.
Zur Zuordnung der Versuchsparameter und Ergebnisse in die verschiedenen Frettingregime wurden Frettingkarten erstellt. Diese erlauben, auf

Basis der Kurzzeitexperimente und der Einordnung in Frettingregime stabile und sichere Betriebsbedingungen vorherzusagen und damit design-relevante Vorschläge zu unterbreiten. Die Kombination einer hoch angelegten Last (z.B. > 75 N) und einer kurzen Amplitude der Gleitbewegung (z.B. < 15 μm), zusammen mit adequaten Gegenmaßnahmen bezüglich der Korrosion (GESA-T91) gewährleisten einen über die Betriebsdauer tolerierbaren Frettingverschleiß.

Abstract

The development of structural materials able to withstand the operating conditions is one of the key issues for the realization of advanced lead/lead alloys cooled nuclear systems (e.g. LFRs and ADSs). Great efforts were done to investigate and improve the resistance of candidate materials against the corrosion mechanisms occurring in liquid lead alloys, which are dissolution of alloying elements and/or extensive oxidation. Slow growing oxide scales, achieved by controlling the environmental condition (i.e. temperature and oxygen content) and/or the material composition (e.g. aluminium surface alloying), can represent a suitable corrosion barrier for long time protection and assure adequate heat transfer (especially important for in-core components).

Corrosion barriers and candidate materials also need to be tested concerning their mechanical and wear resistance properties. One specific challenge for fuel claddings and heat exchanger tubes is the fretting by flow induced vibrations. This particular type of wear can weaken or damage the protective oxide scale promoting dissolution attack and reducing the components lifetime. Thus, investigation, assessment and prediction of the fretting damage are important for design indications, relevant for economics and safety related aspects.

For this purpose, a dedicated facility named FRETHME (FREtting corrosion Test in Heavy liquid MEtals) was designed and realized to simulate, under reactor relevant conditions, the fretting wear at possible friction contacts. In this facility, the fretting action is re-produced by coupling fixed counter-specimens with sliding specimens. The main fretting affecting parameters like sliding amplitude, frequency, applied load and friction force are constantly controlled and monitored during the tests. Temperature and oxygen content of liquid lead are established and maintained in nuclear reactor relevant ranges by a dedicated heating system and gas supply apparatus.

The FRETHME facility was used to carry out an experimental campaign of fretting test in liquid lead aimed to confirm the wear mechanism, investigate the influence of the main affecting parameters on the fretting process and determine the advisable operating conditions to minimize the fretting damage. The interaction of fretting wear and corrosions mechanism (i.e. fretting corrosion) and the role of liquid lead in the wear process were also matter of discussion.

The materials for the experimental campaign were selected among the candidate steels for lead cooled nuclear systems, namely: the austenitic 15-15Ti steel, the ferritic/martensitic T91 steel and the Al surface alloyed T91 steel (GESA-T91).

The tests in liquid lead were preceded by a series of fretting test in air whose results showed general agreement with the available database and attested the reliability and suitability of the test rig to investigate the fretting process.

The experimental outcomes indicate that fretting damage generally increases with number of cycles (or time), with temperature (up to 550°C) and with sliding amplitude. The influence of the applied load is strictly correlated with the sliding amplitude. The fretting damage increases with the applied load up to a turning point and then decreases; the larger the sliding amplitude the higher the value of the applied load corresponding to the turning point. Both applied load and sliding amplitude strongly influence the fretting severity because they both play a central role in determining the fretting regime. The change of fretting regime is proposed to explain the non-monotonic load - fretting wear relation and the non-linear increase of the fretting wear with the amplitude.

Regarding the fretting – corrosion interaction, on the one hand, the corrosion mechanisms affecting steels exposed to liquid lead (i.e. oxidation and dissolution) enhance the fretting wear. However, fretting in liquid lead, due to the lubricated action of the liquid metal, is less severe than in air. On the other end, the fretting action can remove the corrosion barriers (i.e. protective oxide scale and surface alloyed layer) and this favours dissolution attack.

In line with these considerations, GESA-T91 steel, due to the favourable wear and corrosion resistance properties of the surface alloyed layer, showed the best behaviour among the 3 tested materials. On the contrary, due to the high Ni content, 15-15Ti suffers in most of the testing conditions of dissolution enhanced fretting (especially at high temperature); whereas T91 steel, although generally more fretting resistant than 15-15Ti steel, is characterized by oxidation enhanced fretting for temperatures higher than 500°C. According to experimental results, pre-formed oxide scales and Ni-enriched liquid Pb are possible countermeasures to mitigate fretting enhancement by dissolution and oxidation.

Fretting maps aimed to predict fretting severity for a certain set of operating conditions and to give design and assembly indications to minimize the fretting effects were compiled. The combination of high applied loads (e.g. > 75 N) and short sliding amplitudes (e.g. < 15 μm), together with adequate corrosion countermeasures are expected to meet the operating requirements in terms of maximum allowed fretting penetration rate.

Contents

Introduction

A constant increase of energy demand has been recorded for the last decades. The industrial expansion of highly populated countries (e.g. India and China), the technological development and a worldwide technology distribution at every society level are only some of the main reasons which have led to a more and more extensive energy need.
In this context, the prospective of future oil reserve depletion and the necessity of CO_2 emission reduction (Kyoto Protocol) have driven developed countries to an intensive research of energy sources which can be a complement or an alternative to oil.
For electricity production, fossil energy sources (coal, oil and gas), in combination with alternative and renewable solutions (geothermic, hydroelectric, energy provided by the sun and wind) and nuclear energy are considered. In particular, the nuclear power for electricity production has appeared to contribute in a significant way to reduce the CO_2 emission and to develop alternative fuels (e.g. Hydrogen and Synfuel) for transportation and mobility systems.
Since the second half of the XX century, extensive efforts have been made to improve the nuclear power technology. Starting from nuclear systems of the so-called generation I, aimed to demonstrate e.g. the technical feasibility, through generation II and generation III particular emphasis were given to economic and safety related aspects.
Finally, the more recent (2001) Generation IV Nuclear Energy Systems initiative has suggested that, safety, nuclear waste management, proliferation resistance, sustainability, reliability and resource utilization should be pillars of the design and development guideline of the new generation of reactors.
In the framework of the Generation IV International Forum (GIF) to achieve the aforementioned goals different technological solutions are

under investigation. Among these, heavy liquid metals (HLMs) cooled systems are considered.

HLMs, for nuclear application, were studied since the early development of fission energy but more recently a renewed interest in such coolants, in particular for lead and LBE (lead-bismuth eutectic) has been shown. Lead and LBE are coolants investigated for Generation IV LFR (Lead cooled Fast Reactor) and ADS (Accelerator Driven System).

In March 2011, the Fukushima events put under the spotlight of the public opinion the safety of nuclear power plants and the necessity of partitioning and transmutation of nuclear waste.

From this point of view, both LFRs and ADSs are characterized by very high intrinsic safety and they can be used for burning or transmutating minor actinides (MA).

Structural materials development is a key issue for the realization of advanced lead cooled nuclear systems (ADSs and LFRs). Corrosion resistant materials, able to withstand the operating conditions (e.g. temperature and irradiation damage) and to meet the performance criteria are required for structural and in-core applications.

Accordingly, in the last years, substantial efforts were made to investigate and improve the resistance of candidate steels against dissolution attack (loss of alloying elements) and extensive oxidation, which are the typical corrosion mechanisms occurring in liquid lead/lead alloys.

It was found that, if the amount of dissolved oxygen inside liquid lead is adequate, a Fe/Fe-Cr oxide scale develops on the steel surface providing protection towards dissolution. This is one of the main approaches to decrease the corrosion losses and it can be provided by injecting oxygen into coolant and continuously controlling its concentration. However, at temperature above 500°C, even if the oxygen potential in the liquid metal is kept at a proper level, it has been shown that austenitic steels suffer from severe dissolution attack, while martensitic steels are characterized by thick oxide scales that can lower the heat transfer and may break off and eventually plug the coolant channels. Thus, components exposed to high temperature (above 500°C) have to be protected by a suitable surface modification.

A well understood way to provide steel protection at high temperature is the Al-containing surface alloying. Several researches highlighted that surfaces alloyed layers containing Al (e.g. FeCrAlY) have a great potential in protecting steel surface against dissolution and severe oxidation in lead. If the Al concentration is adequate (typically in the range 5-15 wt%), a thin but very resistant alumina scale (Al_2O_3) forms due to Al diffusion to the surface and to selective oxidation. This controlled oxide scale growth showed promising results with respect of preservation of the physical and mechanical properties of the steels. However, such barriers need also to be tested concerning their irradiation resistance and mechanical stability.

One particular challenge which is of interest for specific materials of reactor core and steam generator (SG) is the so called fretting corrosion; a joined action of mechanical wear (fretting) and corrosive environment (liquid lead/lead alloys).

Due to flow induced vibrations (FIV), fuel pins in the reactor core and heat exchanger (HX) tubes in the SG are expected to experience small oscillatory displacements against holders, supports and spacer grids. According to the designers' prediction, frequency and amplitude of the oscillation are expected to reach values up to 40 Hz and 200 μm, respectively.

The resulting mechanical wear can interact with the corrosion processes in liquid lead and weaken or damage the corrosion barriers (e.g. protective oxide scales). Thus, fretting corrosion might reduce the components lifetime promoting dissolution attack and favouring oxide scale break off, which can result in plugging of coolant channels. As a consequence, investigation, assessment and prediction of the fretting corrosion damage are of relevance for safety and economy related issues. Thus, for the final design of lead cooled nuclear systems an overall characterization of fretting corrosion in heavy liquid metals might be also necessary.

Over the years, fretting corrosion and the main variables affecting this process were substantially investigated in air and aqueous environments.

However, only very limited data of fretting tests in lead alloys (not described in all experimental details) are available.
From this point of view, the present work was initiated in the framework of the FP7 GETMAT project (Generation IV and Transmutation MATerials) with the objective to generate design relevant data on the fretting corrosion behaviour of structural steels in liquid lead and lead alloys.
For this purpose, a dedicated facility, named FRETHME (FREtting corrosion Test in Heavy liquid Metals), was designed and assembled to perform fretting corrosion tests in reactor relevant conditions.
The FRETHME facility is a novel and original facility which is used to execute fretting tests at high temperature (up to 700°C), controlling the oxygen content in the liquid metal/alloy and varying the main fretting affecting parameters like sliding amplitude, load and oscillating frequency in ranges of interest.
In agreement with the GETMAT project guideline, the materials under investigation in this work were selected among the main candidates for in-core and in steam generator applications for lead cooled nuclear systems. In particular, the ferritic-martensitic T91 steel, the austenitic 15-15Ti steel and an Al surface alloyed T91 (GESA-T91) were submitted to fretting tests in liquid lead.
Before the experimental campaign in liquid lead, a preliminary series of fretting tests in air was accomplished. These tests were aimed to confirm the wear process, to evaluate the FRETHME facility reliability, to outline the fretting damage assessment route and to create a comparison term with the existing database and with tests performed later on in liquid lead.
This work is organized as follows. In chapter 1, a description of the main lead/lead alloys cooled systems proposed in the framework of the GIF is given and, since this thesis deals with fretting in liquid lead, the attention is particularly focused lead properties and corrosion mechanisms. Chapter 2 reports a detailed literature survey about the fretting process, its main characteristics and affecting parameters. In chapter 3, the design and the functioning of the FRETHME facility is fully described. Chapter 4 exhibits

the experimental activity description such as specimens preparation, fretting tests plan and analysis methods. Results and discussion of the experimental campaigns carried out in air and in liquid lead are reported in chapter 5 and 6 respectively. Summary and conclusion close the manuscript (chapter 7).
The aim of this thesis is to characterize for the first time the fretting process in liquid lead and its interaction with the corrosion mechanisms affecting steels exposed to molten lead alloys (fretting corrosion). The role of the main affecting parameters and indications about possible countermeasures against fretting and fretting corrosion are also objectives of this work. A further target set by the author is the use of the experimental results to compile fretting maps (based on the concept of fretting regimes) to be adopted as tool for fretting damage prediction and for design guideline indication.

1. Lead cooled nuclear systems

1.0 Introduction

The Lead cooled Fast reactor (LFR) together with other innovative nuclear reactor concepts, such as SFR (Sodium Fast Reactor), MSR (Molten Salt Reactor), SCWR (Super Critical Water Reactor), VHTR (Very High Temperature Reactor) and GFR (Gas Fast Reactor), was chosen at the beginning of this century for an international development program in the framework of the Generation IV International Forum (GIF).

The mentioned nuclear technologies were rated as the most promising to satisfy the goals of safety, nuclear wastes management, sustainability, reliability, resources utilization and proliferation resistance; identified as pillars on which to base the future nuclear energy production [1].

Heavy liquid metals (HLM), such as lead (Pb) and lead-bismuth eutectic (LBE), were already matter of investigation as coolants for fast reactors at the dawn of the fission energy. However, in the 1960s sodium became the preferred choice, due to the higher power density achievable with this coolant that resulted in lower doubling times, an important objective during those years of breeding reactors development [2]. Despite of this, Pb and LBE continued to be subject of extensive research activities, mostly in the former Soviet Union where LBE was selected as coolant for alpha class submarines. More recently, around the end of the last century, lead and LBE features and the associated technologies inspired several projects in the emerging field of Accelerator Driven Systems (ADSs); in particular, since the mid of the 1990s, Pb and LBE have been associated as coolant and neutron spallation target for the external neutron source proposed in the ADS concept [2].

The HLM nuclear cooled systems presently under investigation are the so-called LFRs (Lead cooled Fast Rectors) in the GIF framework and ADSs.

Fast reactors are fission reactors in which the neutron spectrum in the core is “fast”, that means that the neutrons are not thermalized (slowed down) as in a conventional water cooled reactor [3]. The fast neutron spectrum is the essential condition for enabling the breeding concept in nuclear reactors and for reprocessing the nuclear waste coming from conventional reactors.
Following the guideline of the GIF, thanks to the Pb and LBE properties, HLM cooled nuclear systems have been associated to different purposes including hydrogen production, nuclear waste transmutation (e.g. by ADS) and small modular reactors with long life cores for supplying electricity/heat in remote areas and developing economies [1, 4].

1.1 Liquid Pb as nuclear coolant - advantages and drawbacks

Besides thermal properties and its compatibility with the structural materials, specific characteristics strictly related to the reactor environment (i.e. neutronics and irradiation effects) must be considered for the selection of nuclear reactor coolants [5].
In a nuclear reactor, the function of the coolant essentially consists in extracting the heat produced by the fission chain reaction occurring in the nuclear fuel elements. Thus, high thermal conductivity and heat capacity are preferable to ensure an efficient heat exchange and transport from the reactor core to the heat exchangers (HXs). Liquid lead is characterized by good heat capacity, excellent thermal conductivity (20-30 times higher than water) and high density.
In Tab. 1.1, the main physical properties of liquid Pb, compared to water and Helium (other nuclear coolants), are reported [6, 7].
The nuclear reactor coolant, like the other elements of the reactor design (i.e. structural and in-core components), interacts with the neutrons released during the fission reactions. Such interaction can be of different nature (e.g. scattering or absorption), depending on the physical properties

of the coolant (e.g. atomic mass), and it influences the reactor core physics, in particular the neutronics. Therefore, the choice of the coolant depends on the purpose and characteristics of the technology selected for the nuclear system [5-8].

	m.u.	Pb	Water	Helium
Atomic mass	$g \cdot mol^{-1}$	207	18	2
Density (at 20°C)	$g \cdot cm^{-3}$	11,34	1	0,17
Melting Temperature	K	600	273	- 272
Boiling Temperature	K	2013	373	-268
ΔV (at melting)	/	3.7		
Heat capacity	$J \cdot kg^{-1} \cdot K^{-1}$	147	4186	5190
Thermal conductivity	$W \cdot m^{-1} \cdot K^{-1}$	15.4	0.6	0.152
Reactivity with air/water	/	low	inert	inert
σ_s (scattering cross-section)	barn	6.4	3.5	3.7
σ_a (absorption cross-section - 1 MeV)	mbarn	6	0.10	0,007
Relative moderating power	/	1	421	0.27

Tab. 1.1

Physical properties of interest for nuclear application of Pb, water and He [6, 7].

Due to its high atomic mass, high scattering cross-section and low absorption cross-section (σ_s and σ_a in Tab. 1.1), Pb shields effectively γ-rays, behaves as a neutron reflector, has low neutrons absorption and scarce moderation (neutron energy loss). The low moderating power (slowdown of neutrons due to energy exchange) and the low neutron absorption allow to operate with a fast neutron spectrum [5, 6, 9]. These features make Pb a promising coolant for next generation fast nuclear reactors (e.g. LFR).
Although a metal, Pb has relatively low melting temperature, high boiling point and does not react readily with air, water/steam, or carbon dioxide [10, 11]. These features bring some important technological advantages like: (i) possibility to operate in ambient pressure, which simplifies design and eliminates the risk of boiling or flashing in case of pressure reduction; (ii) no vigorous exothermic reactions (like Sodium); (iii) no need of intermediate coolant (like in SFRs) reducing deployment and maintenance costs [1]. These features enhance safety and economics aspects. Additional ad-

vantages that increase significantly the safety and that derive from the use of liquid Pb as nuclear coolant are high capability of trapping fission products [12] and high level of natural circulation of the coolant [1]. In fact, the reactor can be designed (e.g. regulating the coolant free level) to ensure natural coolant circulation and heat removal from the core even in case of leakage of the reactor vessel or an unexpected pump breakdown. Besides, due to natural convection, the high density and the high solubility of the actinides in liquid Pb, nuclear fuel dispersion dominates on fuel compaction, avoiding severe re-criticality in accidental situations [1]. These properties together with the decay heat removal (DHR) favored by natural coolant circulation allow less stringent operations timing and high level of containment, which are important aspects in case of accidental situations e.g. analogous to the one of Fukushima nuclear power plant where the cooling system was seriously damaged by the tsunami hit.

Alongside with the several advantages, the choice of liquid Pb as nuclear coolant implies some drawbacks and constrains. Pb is corroding and embrittling most of today available structural materials (Fe and Ni based alloys) [6]. As consequence, the development of protecting oxide scales and adequate corrosion barriers is necessary [13, 14]. In this respect, implementation of adequate oxygen control systems and R&D of protecting methods or alternative alloys are currently under investigation [1, 6, 15]. The low thermal conductivity through the protective oxide scales/corrosion barrier covering the steel surface, the erosion problem and the high Pb density allow only reduced flow rates, which limits the total reactor power [11]. Moreover, the interaction with the structural steel might lead to high content of impurities in the coolant. Additionally, due to the Pb opacity difficulties in inspection/routine operations might arise.

Although the activation of liquid Pb resulting from the irradiation is less significant than for LBE, radiotoxic α-emitting ^{210}Po isotopes can possibly form from Pb according to the following reactions:

$$^{208}Pb + n \rightarrow {}^{209}Bi + n \rightarrow {}^{210}Bi \xrightarrow{\beta\ (5\ days)} {}^{210}Po \xrightarrow{\alpha\ (138{,}4\ days)} {}^{206}Po_{(stable)}$$

^{209}Bi is produced from ^{208}Pb, and ^{210}Po results from a further neutron capture of ^{209}Bi; however the Po inventory is much lower (1 g is expected after 40 years irradiation) than in the case of LBE used as coolant [9].

1.2 Lead cooled Fast Reactor (LFR) concepts

The main features of the LFR systems are: use of Pb or LBE as coolant, the fast neutron spectrum and closed fuel cycle for conversion of fertile uranium (breeding concept) with possibility to transmute the minor actinides of the spent fuel [1, 12]. In 2004, the Lead Fast Reactor (LFR) Provisional System Steering Committee (PSSC) was organized to develop the LFR System Research Plan (SRP). The committee selected two pool-type reactor concepts as candidates for international cooperation and joint development in the GIF framework: the Small Secure Transportable Autonomous Reactor (SSTAR), a 10-100 MWe natural circulation reactor concept with shippable vessel and the European Lead-cooled System (ELSY), a large-sized system rated at about 600 MWe for central station power generation [9].

The peculiarities of the SSTAR are the 20-30 years lifetime cartridge sealed reactor core and system transportability. The sketch of the SSTAR is reported in Fig. 1.1 (adapted from [9]) while the operating parameters and technological characteristics are listed in Tab. 1.2.

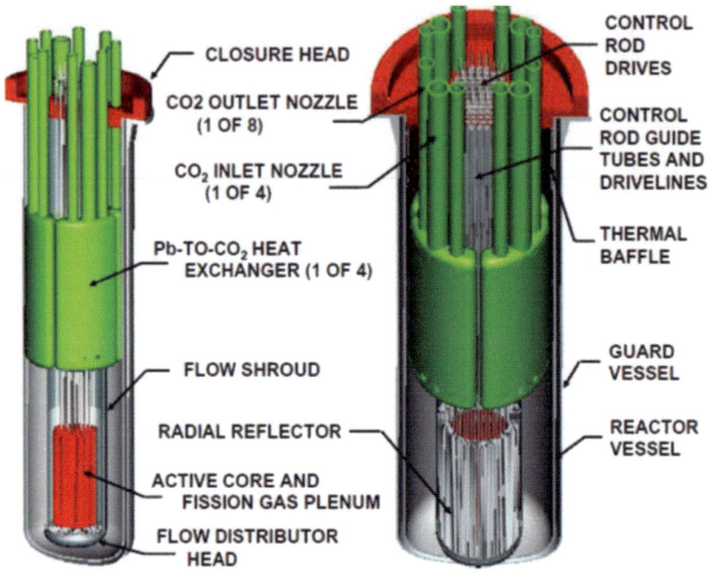

Fig.1.1. Sketch of the Small Secure Transportable Autonomous Reactor (SSTAR) (adapted [9]).

The sealed core avoids or limits the on-site refueling, also considering that once loaded, the SSTAR is fissile self-sufficient since it operates at a conversion ratio equal to 1. The initial fissile inventory for the one-time load is relatively large but substantially smaller than the lifetime fissile consumption (^{235}U) of a LWR for the same energy delivery [4]. The minor actinides created from the nuclear fuel, being securely stored in the sealed reactor core, are not transmuted in other systems (e.g. advanced recycle reactors). As consequence, the SSTAR gives an alternative to nuclear waste management. These characteristics together with the use of liquid Pb as coolant, the design as natural circulation pool-type reactor and the possibility to use nitride fuel, impart to the SSTAR strong proliferation resistance [1, 4]. Besides, the not-access to the fuel limits the operators support requirement.

According to the designers, in the SSTAR the conversion of the reactor core thermal power into electricity is accomplished by a supercritical carbon dioxide (S-CO_2) Brayton cycle. Such solution should lead to a thermal plant efficiency of 44% with small turbomachinery that reduces the cost of the power converter [4, 9].

	ELSY	SSTAR
Power (MWth)	1500	45
Thermal efficiency (%)	42	44
Primary coolant	Pb	Pb
Primary coolant circulation(at power)	Forced	Natural
Core inlet T (°C)	400	420
Core outlet T (°C)	480	567
Fuel	MOX (nitrides)	Nitrides, with ^{15}N
Peak cladding T (°C)	550	650
Fuel pin diameter (mm)	10.5	25
Power conversion system (working fluid)	Water-superheated steam (18 MPa, 450 °C)	Supercritical CO_2 at 20 MPa, 552 °C
Primary/secondary heat transfer system	Eight Pb-to-H_2O SGs	Four Pb-to-CO_2 HXs

Tab.1.2

Main characteristics of the SSTAR and ELSY [9].

Due to the small dimensions (typically the vessel is about 15x3Ø m) and relatively low weight (maximum 500 tons) the entire reactor can be delivered by ship or overland transport, making possible the deployment of such technology in remote areas with minimal industrial infrastructures (e.g. Alaska or Hawaii) [4].

Moreover, the small size of the SSTAR allows to overcome one of the main issues of the designers: to engineer structures able to accommodate the mechanical loads, especially in seismic areas.
Nitride fuel qualification, in-service inspection for components immersed in liquid Pb and R&D of materials able to meet the service requirements are some of the main open issues for realization of the SSTAR.
Concerning the large-sized systems, the results achieved in the framework of ELSY where used as starting point for the European Advances Demonstration Reactor (LEADER) project. The LEADER project has the goal to finalize the design of a large size LFR and to develop the conceptual design of a down-scaled demonstrator [9].
The main objective of ELSY is to demonstrate the possibility of designing a competitive fast critical reactor using simple engineering solutions and complying with the Generation IV goals of: safety, sustainability, proliferation resistance, economics and physical protection [16, 17].
The favorable properties of liquid Pb as nuclear coolant (i.e. low neutron absorption and moderating power) allow a fast neutron flux even with a large amount of coolant in the core, ensuring an efficient utilization of excess neutrons and the reduction of specific uranium consumption. Breeding ratio of about 1, long core life and high fuel burn-up can be also achieved. The fast neutron flux can significantly reduce the nuclear waste generation and LFR systems are capable to safely burn recycled minor actinide within the fuel [3]. The characteristics described so far favor the resource utilization (e.g. Uranium), waste management (e.g. minor actinides containing fuels can be adopted) and proliferation resistance (i.e. breeding ratio equal to 1 and use of liquid metal as coolant).
Many design simplifications derive from the use of liquid Pb as coolant. The pool-type, low-pressure primary system allows the adoption of simple fuel handling systems [1]. The non-reactivity of Pb with air/water/CO_2 [1] enables the use of in-vessel steam generator units and consequent elimina-

tion of intermediate circuit and loops (typical of sodium technology) with gain of around 50% in the reactor volume [18]. Considering that no intermediate circuit degrades the thermal cycle and the relative high expected inlet temperature (~ 400°C), a high-efficiency supercritical water-steam cycle is possible [19]. Additionally, a supercritical carbon dioxide Brayton cycle energy conversion can be considered as alternative solution [1].

The ELSY pool-type reactor concept is shown in Fig. 1.2 (adapted from [9]).

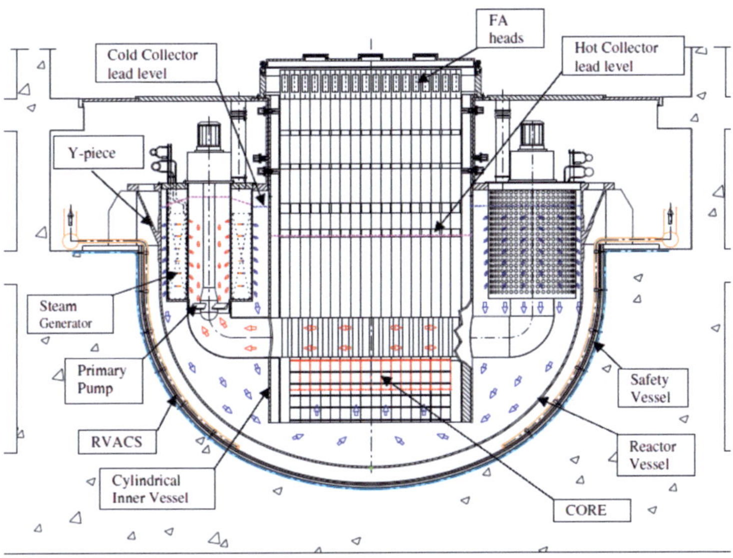

Fig.1.2. Sketch of ELSY and its primary system configuration (adapted from [9]).

In Fig. 1.2, the red and blue arrows display the hot and cold stream of the molten Pb respectively.

To simplify the design, Pb is pumped from the core area, where it warms up by cooling the fuel elements and flows to the HX modules in the SG.

Here, the heat is transferred from Pb to the secondary system coolant (water) and electricity is produced by the typical Rankine cycle. At this point, becoming cold, liquid Pb goes down again to the core area. In case of pumps breakdown, natural circulation is sufficient to ensure the decay heat removal.

Both the immersed pump impeller and the spiral once-through SG illustrated in Fig. 1.3 (adapted from [9]) are examples of the innovative solutions adopted in the ELSY design [9].

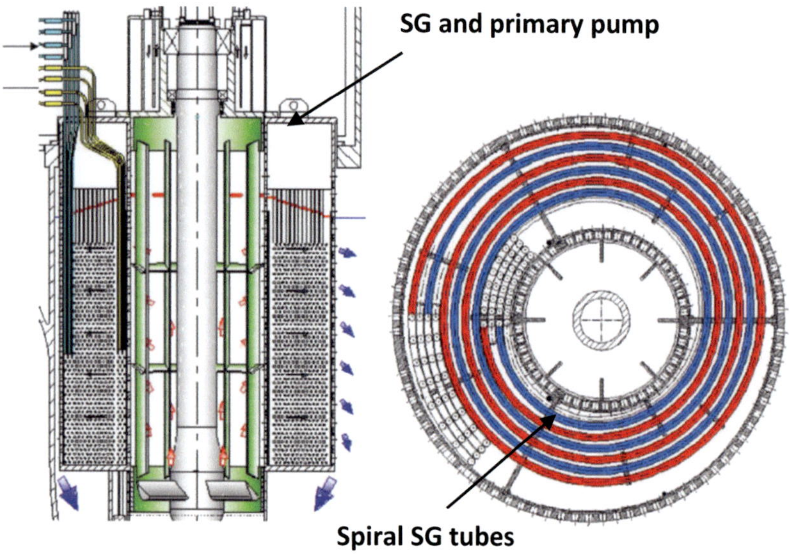

Fig.1.3. Steam generator, primary pump and spiral SG tube (adapted from [9]).

Structural materials stability plays a central role in the choice of the inlet/outlet temperature, which are 400 and 480°C respectively. This operating temperature range, which still ensures a sufficient margin from Pb freezing, offers some advantages compared to higher temperature ranges, like: (i) lower embrittlement of the structural materials by liquid Pb and fast neutron flux; (ii) reduced thermal stresses and (iii) mitigation of corrosion and creep effects. From another point of view, a thermal cycle with larger range of temperature would be desirable because it would allow lower liquid Pb flowing rate (resulting in less erosion), further compactness and the possibility to increase the reactor power [9]. ELSY, in comparison to other and more complex types of nuclear reactors, is characterized by significant design simplifications aimed to reduce the general capital cost that is the major cost factor for the competitive generation of nuclear electricity [1, 9, 18, 19]. Thus, it can be assumed that the lead cooled technology takes into account another of the main pillars of the GIF guideline: the economics.

1.3 Accelerator Driven System (ADS)

In the strategy of the world nuclear power development, a closed nuclear fuel cycle is necessary for efficient resources utilization and nuclear waste minimization. In this respect, Partitioning and Transmutation (P&T) techniques, involving the use of Accelerator Driven Systems (ADSs) in the nuclear fuel cycle, are necessary for the burn-up and transmutation of radioactive wastes created in conventional reactors [20-21].

The concept of ADS was born in the first half of 1990s by a group of researchers led by Carlo Rubbia [22]. Accelerator Driven Systems are nuclear fission reactors with a sub-critical core. In other words, ADSs cannot establish a self-sustaining chain reaction to produce fission because criticality cannot be achieved. Therefore, such a system requires an external neutron source for establishing a stationary behavior of the core; for this purpose most of the proposals assume the use proton accelerators [3]. As shown in Fig. 1.4 (adapted from [23]), an ADS consists of three main parts: an accelerator for generating protons; a spallation target (or target module), to produce free nucleons (neutrons) by the protons in spallation reactions (external neutron source); a subcritical blanket (the core of the ADS) [22, 23].

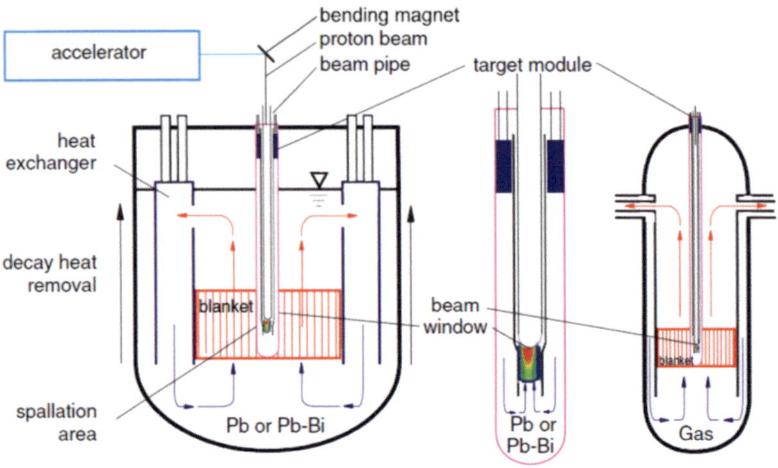

Fig.1.4. Sketch of ADS system, both the HLM cooling and gas cooling possibility are displayed (adapted from [23]).

The external neutrons escape from the spallation target and induce fission reactions in the blanket region that is similar to a conventional reactor core. By the fission reactions, additional neutrons (inside source) are released in a non self-sustaining fission chain together with thermal energy. These internal neutrons convert fertile isotopes into fissile ones, which is the concept of transmutation [6]. The external neutrons produced in the spallation target are completely independent from the subcritical blanket. A shutdown of the accelerator or an interruption of the proton beam immediately stops the fission reaction, except for a small fraction originating from delayed fission neutrons [6, 22].

ADS can be designed to operate either with a thermal or fast spectrum [24, 25]. Two possible options regarding the coolant have been considered: on one side, HLMs such as LBE and Pb, on the other side a gas-cooled system. In Fig. 1.4 a sketch of both options are shown [23].

The spallation target has to provide the highest possible neutron yield and has to be transparent to neutrons. From this point of view, a possible spallation target is provided by a heavy liquid metal target, typically Pb or LBE, since they have excellent nuclear features and they can participate in cooling [26]. A high intensity (~mA) beam of protons is accelerated to high energy (hundreds of MeV to GeV), hits the HLM target, Pb or LBE, and generates, very efficiently, neutrons in the liquid metal/alloy by spallation reactions (with a yield of about 30 neutrons per incident proton of 1 GeV in Pb). Thus, the neutrons produced sustain (*drive*) the chain reaction in the sub-critical core. This fission reaction in the fuel region is necessary to transmute long lived isotopes (mixed in the fuel elements) to short lived ones and to allow the criticality of the system to be sustained and controlled [2, 3, 23].

Because of its characteristics, the ADS concept complies with the goals of safety (i.e. sub-criticality and fast shut down) and nuclear waste management (ADS can be designed as transuranium elements or minor actinides burner – TRU or MA burner) highlighted in the GIF framework [3].

Of course many open issues are still associated with this technology, i.e.: containment questions related to the interfacing accelerator/reactor; the target and the surrounding structural material are subjected to complex degradation phenomena due to an interaction of thermo-mechanical loads, high energy particles irradiation and corrosion effects due to the HLM [2, 3, 21].

In the EURATOM framework, the EUROTRANS project had the purpose to demonstrate the feasibility of industrial transmutation, through ADSs, of high level nuclear waste such as minor actinides (MA) [27]. Such objectives were met proposing two technological approaches: an industrial transmutation plant and a demonstrator aimed to validate the technological feasibility of coupling an accelerator, a neutron spallation target and a sub-critical core. The first task was associated to the European Facility for Industrial Transmutation (EFIT), the second to the European Experimental Accelerator Driven System (XT – ADS), which was more recently replaced by the MYRRHA concept (Multi-purpose hYbrid Research Reactor for High-tech Applications).

The coolants are Pb and LBE for the EFIT and MYRRHA respectively, which are also intended to serve as target material for the external neutrons source [28]. Some of the most important features of EFIT and MYRRHA are listed in Tab. 1.3 [27-30].

The working temperatures in EFIT are higher than in MYRRHA; in fact in order to have a sufficient margin above the Pb freezing point (327°C) the inlet temperature is 400°C, whereas in MYRRHA it is 270°C (freezing point of LBE is 125°C). Therefore, LBE has the advantage of allowing operation at a lower temperature. From the point of view of neutronics, both HLMs are satisfactory. Concerning the fuel, in EFIT uranium free fuel in considered, it is composed of plutonium and americium oxides in a magnesium matrix (molybdenum is the backup option); for MYRRHA, being a demonstrator system, the proposed fuel is the MOX type with 30-35% of plutonium oxides. Besides, EFIT has the capability to work both as MA burner and for electricity generation [27-30].

	EFIT	MYRRAH
Power (MWth)	~ 400	100
Primary coolant	Pb	LBE
Primary system	Pool-type integrated	Pool-type integrated
Core inlet T (°C)	400	270
Core outlet T (°C)	480	410
Power conversion system working fluid	Water - superheated steam	/
Fuel	$(Pu,Am)O_2$+MgO/Mo	MOX (30-35% PuO_2)
Fuel loading	Top	Bottom

Tab.1.3

Main characteristics of EFIT and MYRRHA [27-30].

1.4 Structural materials for lead cooled nuclear systems

Economics, safety, efficiency, and nuclear waste management (e.g. partitioning and transmutation) are some of the main objectives of the technological guideline for GEN IV fast nuclear reactors. Achieving these goals in an efficient manner requires particular operating conditions, generally more stringent than the presently operating reactors, such as higher burn-ups (up to 3-4 times higher), longer irradiation of the fuel elements, higher working temperature and significantly harder neutron flux. In this context, the development of clad and structural materials able to withstand the operating conditions is one of the key issues for the feasibility and deployment of the next generation advanced nuclear technologies.

There are many requirements for nuclear reactor materials, regardless of the particular type of reactor for which they are considered. Availability, costs, easy fabrication and joining properties are only few of the most general considerations that should be made selecting materials for nuclear power plants [31]. In the specific case of the clad and in-core components, due to the high operating temperatures, the materials must have good mechanical properties at elevated temperature, including creep resistance, long term stability and compatibility with the reactor coolant. Moreover, the hard neutron flux (up to 10^{15} $n \cdot cm^{-2}s^{-1}$) and the expected longer permanence in-core (at least 3 years for the fuel clad) is expected to cause significant irradiation damage (greater than 200 dpa). As a consequence, resistance to irradiation-induced property changes, such as hardening, embrittlement/helium-induced embrittlement, swelling, creep and phase change/instabilities, is required. Good neutronics (low absorption of neutrons) and negligible fuel/clad chemical interaction are also important factors, especially for clad materials.

For structural components like the core vessel and supports for fast nuclear reactors, material requirements and challenges similar to the in-core components can be associated. Furthermore, structural components are expected to serve the entire operating life of the nuclear system, estimated around 60 years for commercial reactors. Thus, for these components, more importance is given to creep resistance, thermal stability and fatigue properties. The neutron fluencies in the area of structural parts are considerably lower than for the in-core component so that here also the irradiation damage is lower: 5-10 dpa instead of up to 200 dpa. However, due to the longer exposure, also for these components, the irradiation influence on the materials properties should be carefully considered [32].

In the specific case of the lead cooled nuclear systems, one of the most critical drawbacks related to the use of liquid Pb as nuclear coolant is its aggressiveness towards structural materials, in terms of corrosion, erosion and embrittlement. As consequence, together with the typical steel features required for nuclear applications (mechanical properties and dimensional stability over the operating of temperature, stresses and doses), corrosion, erosion, stress corrosion and liquid metal embrittlement resistance are also required.

The main materials currently under investigation for lead cooled nuclear systems are listed in Tab. 1.4.

Material	Cr	Ni	Mo	Ti	Si	C	V	Mn	W	Y_2O_3
T91	8.3-8.6	-	0.9-0.95	-	0.3-0.4	0.11	0.2	0.4-0.7	-	-
T92	8.5-9.5	0.4	0.3-0.6	-	0.5	0.09	0.04	0.3-0.6	1.5-2	-
ODS 9	9	-	1	-	-	-	-	-	-	0.35
ODS 12	12	-	-	0.29	-	-	-	-	2	0.2-0.25
ODS 14	13.5		0.3	0.26	0.3	-	-	-	1	0.3
316 L	16-18	10-17	2-2.75	-	0.1-1	0.01	-	0.2-2	-	-
15-15Ti	14.7	14.7	1.6	0.43	0.43	0.46	-	-	-	-
MAXTHAL			Ti_3SiC_2							

Tab.1.4

Principal candidate materials for lead cooled nuclear systems with their relative composition expressed in wt%. In the table Fe is balanced.

As aforementioned, the selection of the materials is strongly connected to the area of use in the nuclear reactor. Austenitic steels, especially the low carbon grade like the AISI 316 L, as already proved in the previous generation nuclear reactors (e.g. LWR), are suitable candidates for components operating at low temperature and irradiation damage, like reactor vessel, refueling equipment and purification system.

Due to microstructural stability, swelling resistance and adequate mechanical properties after prolonged irradiation, the ferritic-martensitic steels (e.g. 9Cr1Mo steel T91), are proposed for components subjected to the most severe conditions (high thermo-mechanical load and irradiation flux) such as: in-core components (e.g. fuel cladding, holders and spacer grids), steam generator, heat exchanger tubes and spallation target. For this applications domain, alternative solutions have been recently proposed, like ODS (oxide dispersion strengthened) steels, in which a fine distribution of Y_2O_3 (or TiO_2) particles mechanically driven into the alloy enhances the material properties at high temperature, and ceramic materials like MAXTHAL (Ti_3SiC_2) proposed for the ELSY pump impeller due to its high erosion resistance.

Materials nowadays under investigation like T91 or ODS steel, which are the main candidates for applications in future industrial nuclear power plant (e.g. ELSY), need to be qualified and licensed as nuclear grade in-core materials. Such procedure might take 10-15 years. Thus, in the design of the demonstrator reactors, i.e. ALFRED (Advanced Lead Fast Reactor European Demonstrator) for the lead cooled fast reactor and MYRRHA (Multi-purpose hYbrd Research Reactor for High-tech Applications) for the LBE cooled ADS, the used of the austenitic steel 1.4970 (15-15Ti) is foreseen also for in-core components [27-32]. Indeed, a fast realization of the demonstrators, achievable only using materials already licensed (i.e. 15-15Ti), can give the possibility to qualify new materials (i.e. T91 or ODS steel) in reactor conditions. In this way, the deployment timing for future Pb/Pb alloys cooled systems, scheduled around the year 2050, might be respected.

1.5 Corrosion and protection of steels exposed to liquid lead

As anticipated in paragraph 1.1, one of the main issues related to the use of liquid Pb as nuclear coolant is its aggressiveness towards candidate cladding and structural steels. In particular, such materials in contact with liquid Pb are susceptible to dissolution, a corrosion mechanism that causes the loss of alloying elements [33]. Such process takes place due to high solubility in Pb of the main steel alloying elements. In Fig. 1.5, solubility in liquid Pb of Ni, Al, Cr and Fe, as function of the temperature, is reported. Among these, Ni is the alloying element with the highest solubility [11, 34].

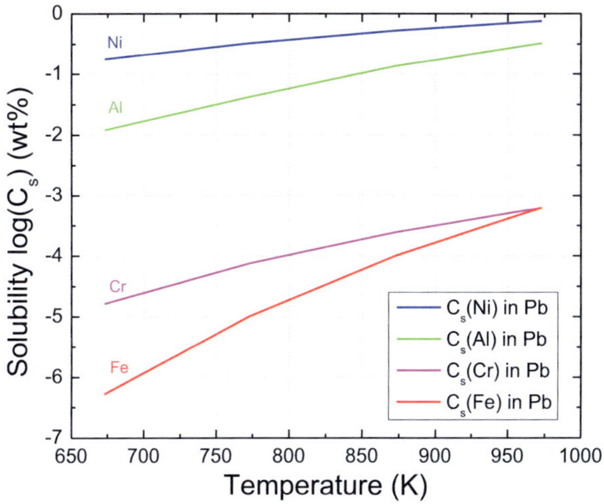

Fig.1.5. Solubility in liquid Pb of the steel main alloying elements.

A promising countermeasure adopted to mitigate dissolution attack occurring in liquid Pb, consists in exploiting the oxidation process, another corrosion mechanism that might occur in liquid Pb due to the solubility of oxygen in liquid Pb (Fig.1.6). Controlling the oxygen content in liquid Pb, it is possible to favor the formation of a thin protective oxide scale. Due to the low diffusion rate of the steel alloying elements through oxide scales, the dissolution of such elements is hindered [33, 35].

The control of the amount of oxygen in liquid Pb plays a key role for in determining the formation of an adequate protective oxide scale on the steel surface and in avoiding extensive oxidation, which might lead to serious drawbacks for the systems and components performance.

Extensive research activities has been carried to investigate and characterize oxidation and dissolution processes in Pb and Pb alloys of the main candidate materials for lead alloys cooled nuclear systems.

In the following paragraphs the literature survey about these corrosion mechanisms, for the particular case of the candidate materials for LRSs and ADSs, is reported.

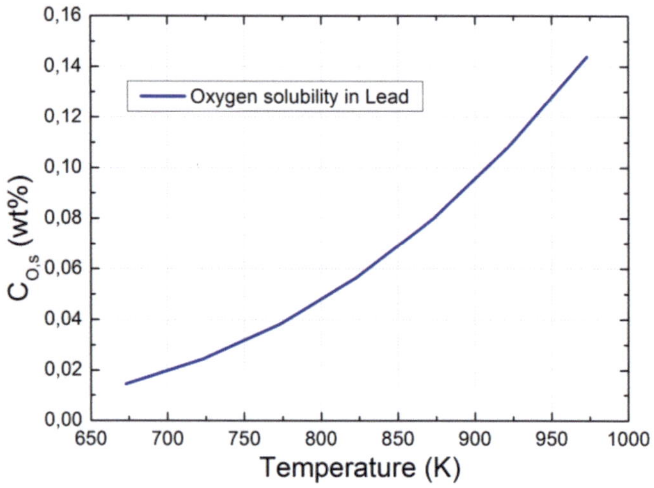

Fig.1.6. Oxygen solubility in liquid Pb as function of the temperature.

1.5.1 Dissolution of the candidate steels for lead cooled nuclear systems

The dissolution process affects steels due to the high solubility in Pb and Pb alloys of the main alloying elements, such as Ni, Cr and Fe. The loss of such elements from the steel matrix influences the components mechanical properties. Besides, considering that solubility is temperature dependent and that the nuclear reactor is not an isothermal system, precipitation of the solutes (Ni, Cr or Fe) might occur in the coldest area of the cooling system, affecting its general efficiency [36].

According to different experimental activities, ferritic-martensitic steels (e.g. 9Cr1Mo T91) are less susceptible to dissolution attack than austenitic steels (e.g. 15-15Ti 1.4970). Indeed, austenitic steels at temperatures above 500°C are affected by dissolution attack even after relatively short time (1000 h), while at temperatures up by about 500°C no dissolution occur even for long term exposure in liquid Pb (10000 h) [37]. In the case of ferritic-martensitic steels, the threshold dissolution/not dissolution is shifted of about 50°C to higher temperatures. 550° C is the border temperature above which dissolution occurs after short/medium term. Factors explaining the different behavior of the two types of steel are e.g. the high Ni content of austenitic steel candidates for nuclear application (15-20 wt %), which among the alloying elements has the highest solubility. Nonetheless, the oxide scale developing on steels exposed to liquid Pb and acting as a protective barrier against dissolution (paragraph 1.5.2), grows faster and it is stable at higher temperatures for ferritic-martensitic steels than for austenitic steels (paragraph 1.5.3).

1.5.2 Oxidation as protecting method against dissolution

Although it is a corrosion mechanism, oxidation can be exploited as protecting method against dissolution. The developing oxide scale, if stable, can provide an effective protection against dissolution, behaving like a corrosion barrier [38-40]. Acting on the steel composition and controlling the oxygen content in the liquid Pb [35], it is possible to favor the development of an adequate thin protective oxide scale passivating the steel surface without affecting significantly the components performance. Due to the low diffusion rate of the steel alloying elements through oxide scales, the direct dissolution is reduced or completely inhibited, it becomes negligible and corrosion proceeds through dissociation or transport through the oxide layer, which is much slower than direct dissolution [41]. The protective oxide scale hinders cations diffusion from the steel bulk to the liquid Pb and vice versa [38].

To develop a protective and ‘‘self-healing’’ oxide scale on the steel surface, the oxygen concentration in liquid Pb must be adequate to favor the oxidation of steel metal cations (Fe, Cr and possibly Al) without causing Pb oxidation. The formation of Pb oxides (e.g. PbO) might induce problems in the cooling channels like plugging due to precipitation. As shown by Eq. 1.1 and in the Ellingham diagram of Fig. 1.7, from the thermodynamic point of view, the free Gibbs energy of formation related to the oxidation process must be lower than the one of formation of PbO and higher than the one for the formation of Fe_3O_4.

Eq.1.1. $$2\Delta G^0_{PbO} > RT\ln p_{O_2} > 0.5\Delta G^0_{Fe_3O_4}$$

These conditions, for the temperature range of interest for lead cooled nuclear systems (400-600°C), occur with an oxygen concentration in the range around 10^{-6} wt% (Fig. 1.8) [10].

As reported in [42] an ideal oxide scales should be free of cracks and non-porous, stress free at the operating temperatures and resistant to spalling-off and heating-cooling cycles. Besides, the growing rate should be low enough (low diffusion of O and Fe/Cr ions) not to cause an excessive recession of the original surface during the envisaged service life [43]. An excessive amount of oxygen in liquid Pb favors extensive oxidation, which will influence the performance and working lifetime of the nuclear reactor components and it also might lead to safety related issues.

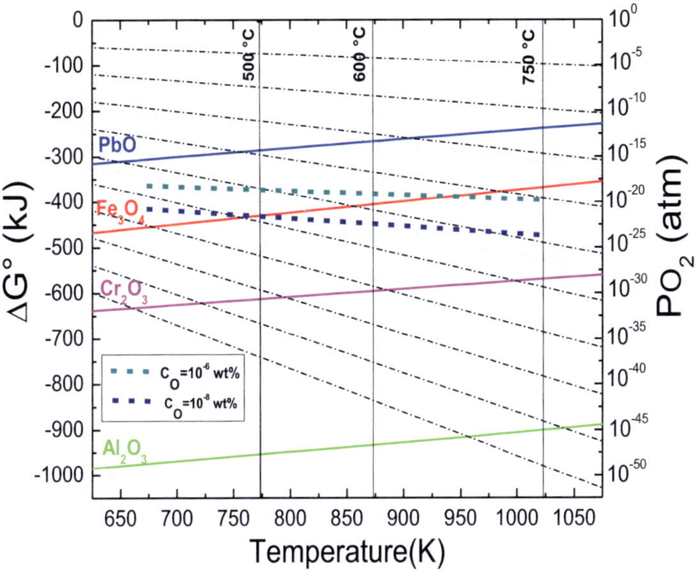

Fig.1.7. Ellingham diagram for Pb, Fe, Cr and Al oxides formation.

The main problems related to the growth of a thick oxide scale on the steel surface are for instance: reduction of the heat transfer due to the low heat conductivity of the oxide scale (e.g. for Fe_3O_4 - 1 $W \cdot m^{-1} \cdot K^{-1}$) and break-off (spall-off) of the oxide scale with risk of cooling channels plugging. In the case of fuel claddings or the heat exchangers (HX) tubes, for which an efficient heat transfer is required, the development of a very thick oxide scale on the steel surface will reduce drastically the heat exchange (i.e. cladding-cooling liquid Pb), leading to overheating of the component and possible occurrence of hot-spots in the reactor core [44]. Besides, as explained in paragraph 1.1, the presence of an oxide scale can influences the design parameters such as the coolant flowing rate [12].

One of the methods which can be adopted to establish the desired oxygen content in the molten Pb is via-control of the oxygen partial pressure in the atmosphere above the liquid metal. A detailed description of this method is given in Chapter 4, where the experimental setup is described.

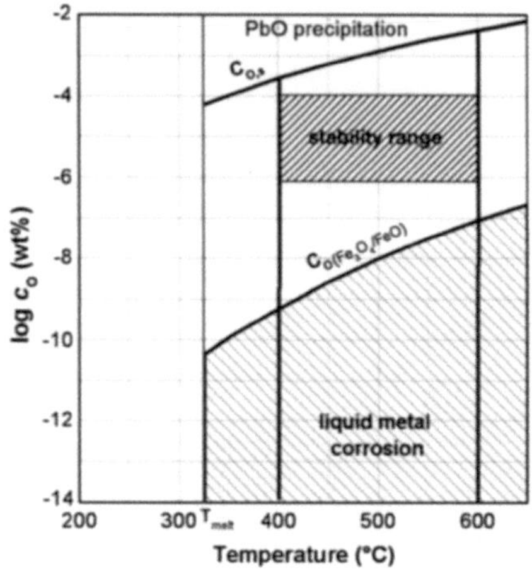

Fig.1.8. Values of oxygen concentration for the formation at different temperature of Fe_3O_4 and PbO (adapted from [10]).

1.5.3 Characteristics of steel oxidation in liquid lead

As reported by Müller et al. [38], there is no principal difference between oxidation in air and in liquid Pb/Pb alloy except for the growth rate of the oxide scale that is smaller in liquid Pb/Pb-alloy; i.e. the growth rate of the magnetite scale (Fe_3O_4) is two to three times lower in HLM than in atmosphere for the same oxygen activity. However both in air and in liquid Pb the oxide scale growth follows a parabolic law [39].

According to the results of several experimental campaigns available in the literature [41, 45, 46], a double-layer structured oxide scale (duplex) seems to be common for steels exposed in liquid Pb containing the sufficient amount of dissolved oxygen. Generally, such scale consists in an outer layer of iron oxides (magnetite, Fe_3O_4) and an inner layer, more compact but less uniform, of Fe-Cr spinel oxide. The magnetite/spinel interface lies at the original liquid/steel interface; indeed the spinel layer grows in the bulk, driven by the oxygen inward diffusion, whereas the magnetite due to the iron cations outwards diffusion grows externally in the liquid Pb (Fig. 1.9 adapted from [47]).

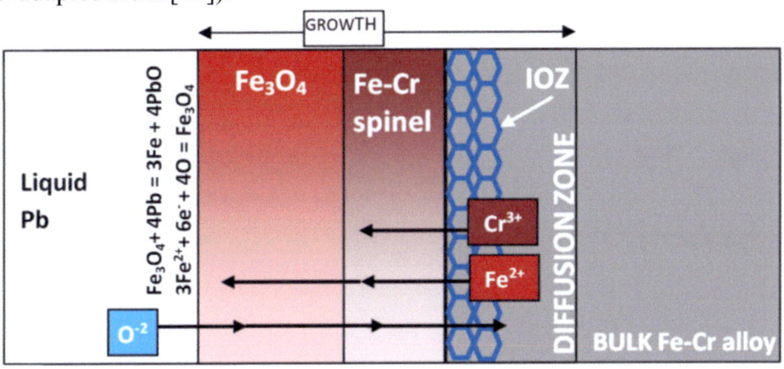

Fig.1.9. Schamatic illustration of the double structured oxide scale forming on steel exposed to liquid Pb and of the diffusion mechanism through the oxide scale (adapted from [47]).

The duplex oxide scale forms due to the different thermodynamic stability and diffusion rate of Cr and Fe, the main steel components. Although in the thermodynamic system Fe-Cr-Pb-O the Cr oxides is the compound with more negative free energy of formation, the growth of a pure chromia scale (Cr_2O_3) is unlikely due to limited amount of alloyed Cr in the steels of interest for nuclear application (< 15 wt%). Cr concentration is nearly zero in the outer layer while it is higher than the nominal composition in the spinel oxide scale. In addition, as schematically shown in Fig. 1.9, beneath the Fe-Cr oxide scale, there is a diffusion region characterized by the Fe cations outwards diffusion (driving the oxidation process) and the inwards diffusion of oxygen through the duplex oxide scale. The cations vacancies created by diffusion processes favor the formation of the so called internal oxidation zone (IOZ). The IOZ is a sort of not yet completed oxidized layer mainly consisting in Cr oxides (Cr_2O_3) generated at the grain boundaries [47].

Since one of the goals of this work is to evaluate the interaction between corrosion mechanisms and mechanical fretting occurring in the fretting corrosion process in liquid Pb, an overview of the oxidation behavior of the steels considered in this work is provided hereafter.

Austenitic steels (e.g. 15-15Ti 1.4970) show better resistance to oxidation than martensitic steel (e.g. 9Cr1Mo T91), which is attributed to the higher Cr content (~15 wt %) [44]. The formation of Cr oxides slowers the outward and inward diffusion of iron and oxygen respectively [45]. Besides, the austenitic structure is characterized by diffusion coefficients up to 4 times lower than the martensitic structure, for all those elements involved in the oxidation process such as Fe and Cr. As a result, oxide scales formed on austenitic steels are generally thinner than those formed on martensitic steels. For the temperature range of interest for the experimental activities of this work (450-550°C), the oxide scale thickness after exposure time < 1000 h is up to about 10 μm and about 15 μm for 15-15Ti and T91, respectively [50-55].

According to the corrosion tests reported in [50-52], at temperature up to 550°C, the ferritic-martensitic steels are covered by a duplex oxide scale; at higher temperature magnetite starts thinning and above 600°C it disappears so that only a Cr rich Fe-Cr spinel oxide scale is visible.

Regarding the austenitic steels, below 500°C a spinel Fe-Cr oxide scale covers the steel surface, around 550°C a duplex oxide scale might be noticed and at higher temperatures no magnetite is visible and the spinel scale is unstable and penetrated by liquid Pb.

1.5.4 Approaches for mitigating corrosion

To reduce steel corrosion (extensive oxidation/dissolution) in liquid Pb, two kinds of countermeasures can be adopted: i) improving the corrosion resistance of the material by e.g. controlling the composition or surface alloying; ii) mitigating the corrosiveness of the liquid Pb/Pb alloy by e.g. addition of inhibitors or controlling oxygen content in liquid Pb. However, the best results in terms of corrosion protection are achieved by working on both of these methods. Indeed, on one hand the control of the oxygen content in the liquid Pb hinders dissolution and avoids extensive oxidation that reduce the components physical (e.g. heat conduction) and mechanical properties; on the other hand an appropriate surface alloying can ensure protection against dissolution by favoring the formation of a thin, protective and slow-growing oxide scale.

The Fe/Fe-Cr oxide scales developing on the candidate steels are stable and protective up to 500°C and 550°C for austenitic and ferritic/martensitic steels respectively; at higher temperature or for long exposure time, excessive thickness, spall-off and lead penetration affects the duplex oxide scales [38, 46]. To ensure the development of a suitable protective oxide scales, stable also at temperatures higher than 500-550°C (occurring e.g. in accidental situations) and for long residence time in molten Pb, one of the most promising solutions is the Al surface alloying. Such surface alloying is obtained by a two steps process: first a LPPS (Low Pressure Plasma Spray) of FeCrAlY powders on the steel surface (austenitic or ferritic/martensitic); second an electron beam treatment (i.e. GESA, Gepulste ElektronenStrahl Anlage), by which the deposited powder is melted with the first tens of micron of the base steel to create the surface alloy. The result of this treatment is a 20-30 μm thick Fe-Cr-Al surface alloy (the thickness depends on the beam intensity) that does not influence significantly the mechanical properties of the components, reduces less significantly than the Fe/Fe-Cr oxides the heat conductivity (10-15 time less) and favors the development of a thin, protective, slow-growing Al oxide scale [38, 53].

However, the protective oxide scales growing on the candidate steels exposed to liquid Pb, both the Fe/Fe-Cr oxides scales and the alumina scale of the Al-alloyed steels, must be characterized also concerning their mechanical properties and irradiation resistance. From this point of view, fretting, a particular kind of wear arising in specific areas of the lead cooled nuclear systems (e.g. claddings and HX tubes) due to flow induced vibration (FIV), might affect the stability of the protective oxide scales and corrosion barriers. Thus, the characterization of this phenomenon is important for determining the components lifetime and performance in operating conditions, which are relevant elements for safety and economics issues. The state of the art of fretting and fretting corrosion together with an analysis of these wear/corrosion processes in the specific case of liquid Pb cooled systems are described in the following chapter (Chapter 2).

2. Fretting – State of the Art

2.0 Introduction

"The forms of damage which arises on steel surfaces in contact and nominally at rest with respect to each other", with this definition in 1927 Tomlinson first coined the term fretting corrosion [56].

Fretting is a particular type of surface damage occurring wherever short amplitude reciprocating sliding between contacting surfaces is sustained for a large number of cycles [57, 58].

Although originally thought to be a kind of corrosion mechanism, due to the powdered oxidized debris produced by fretting when it affects common stainless steels, it became clear that fretting is a much more complicated process which, beside corrosion, might involve adhesion, abrasion, fatigue and can also affect noble metal (e.g. Au and Pt) [56, 59-61].

The range of sliding amplitude in which fretting exists is not well defined. On one hand it is hard to define the lower limit of the range. For instance, reciprocating movements of 0.1 μm of amplitude maintained for a sufficient number of cycles can cause components failure due to fretting related reasons [57]. On the other hand, it is not clear which value of amplitude corresponds to the upper limit that coincides with the transition from fretting to reciprocating sliding wear. Such value is probably a characteristic to be considered case by case [62, 63], but it seems to be in the range 150 - 300 μm [64-68].

The peculiar small sliding amplitude of fretting together with the contact load that characterizes every tribological system dictates the unique features of this particular type of wear [64].

Typical of fretting is the so called third body interaction. The small relative displacement favors the retention in the contact area of the debris (metal and/or oxides) generated during the fretting process. Between the sliding surfaces, the fretting debris can either act as abrasive particles enhancing the surface degradation or they can be compacted in interposing scale that self-limits the wear process. Thus fretting debris can be considered a third body of interaction, which can affect non-negligibly the tribological system [2, 69-71].

Three different types of surface degradation can be associated to the fretting process: fretting corrosion, fretting wear and fretting fatigue [57, 59, 62, 70]. Such mechanisms interact between each other and render the fretting process very complex and comprising of physical, mechanical and environmental factors (up to 50 parameters can affect the fretting process [72]). Therefore, any experimental investigation is very complicated to accomplish. As a consequence, discordant results are often reported in the literature, no universal theory about fretting is available and fretting systems must be often considered as singular system [56-61, 73].

Fretting is of great interest for industrial plants because it can cause premature failures (e.g. by fretting fatigue) reducing components performance and lifetime. Moreover, most of the engineering systems with movable components, where vibrations may induce uncontrolled movements of parts intended to be fixed to each other (e.g. quasi static-assemblies like cables, turbine blade assemblies and nuclear reactor components), suffer of fretting related issues [74-77].

A lack of experimental data characterizes fretting in nuclear power plants for the particular case of future lead alloy nuclear cooled systems. The goal of this thesis is to investigate and characterized fretting wear and corrosion in liquid lead for nuclear reactor components.

To provide a theoretical base for the interpretation of the empirical outcomes of the present work, in the following pages an overview of fretting and of its most peculiar aspects is reported.

2.1 Characteristics of the fretting contact

The fundamental characteristic of fretting is the very small amplitude of sliding (typically below 200 µm [56]), from which the unique features of this type of wear derive.

Applying an increasing tangential force on solids that are pressed together, macroscopic sliding will result if a certain value of this tangential force is passed (gross slip). Although this is a well known experimental fact, it is less widely realized that at levels of tangential force below this limiting value, tangential micro-displacements also occur in response to the applied force.

In other words, under certain conditions of normal and tangential load applied to the contact, a microscopic movement within the contact area takes place even without gross slip. The center might be stationary (unslipped) while the edges of the contact reciprocate with a certain amplitude to cause fretting damage [57].

To explain these micro-movements, discovered to be a characteristic of any Hertzian contact (pure elastic contact between two non-conform bodies) subjected to tangential force [78, 79], an earlier elastic and a more recent elasto-plastic model have been proposed.

In Fig. 2.1 (adapted from [57]), a stationary Hertzian contact characterized by a normal load P applied in the center of the contact, a superimposed tangential force T and a coefficient of static friction μ constant across the contact is represented. The pressure p generated by the applied load P decreases from a maximum value at the center of the contact ($p_{max}=P$) to zero at the contact edges. Similarly, frictional stress $f=\mu \cdot p$ exhibits values from zero at the edges to the maximum in the middle of the contact ($f_{max}=\mu \cdot P$). Besides, tangential stress τ, generated by a tangential force T, reaches from a finite value in the center of the contact to infinite value at the edges. If $T<\mu \cdot P$ no sliding occurs. However, micro-slips might occur at the contact edges where τ arises to an infinite value ($\tau>f$). Thus, considering a constant normal load P and a tangential force T which increases gradually, micro-slips interest immediately the edges of the contact and spread inwards until T approaches $\mu \cdot P$; for higher values ($T > \mu \cdot P$) the contact starts sliding (gross slip) [78, 79].

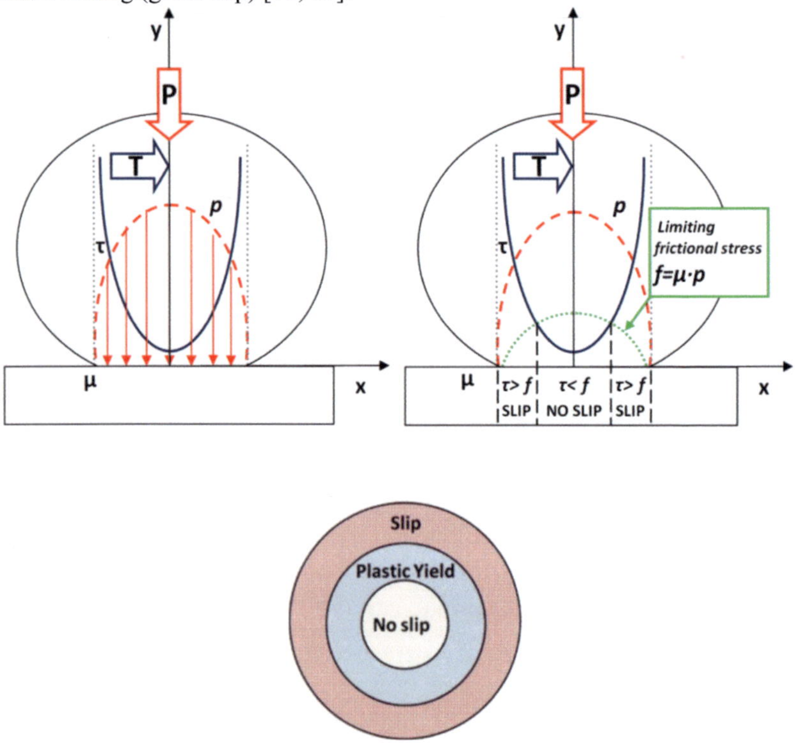

Fig.2.1. Normal and tangential stress fields for Hertzian contact (upper images) and illustration of the elastic and elasto-plastic model (adapted from [57]).

The existence of slip and non-slip regions in Hertzian contacts was confirmed experimentally by testing different sliding amplitudes for constant applied loads to determine at which value of the amplitude gross slip takes place and the unslipped zone stops existing [80, 81].

In the elastic model of fretting, the relative displacement is assumed to be accommodated by microslips between the surfaces and elastic deformation of the contacting solids. Considering the classic theory of friction, the contact occurs as result of rigid contact of the single surfaces asperities. When the shear stress exceeds a critical value ($\tau > f$ in Fig. 2.1), slip occurs resulting from sudden asperities fracture (fretting debris), without any deformation. This simplified assumption is the major difference between the elastic and elasto-plastic model. The latter, compared to the elastic model, predicts higher amplitudes corresponding to incipient gross slip. According to the elasto-plastic model, three areas, differently affected can be identified in the contacting surfaces (Fig. 2.1): (i) a central unslipped area elastically deformed, (ii) an external "slip" zone where the asperities are fractured similarly to the elastic model and (iii) an annular zone between (i) and (ii) in which the asperities were deformed plastically but are not fractured [82].

2.2 Characteristics of the fretting damage

During the fretting process different types of surface degradation and modification might occur. Corrosion, wear, fatigue cracking, abrasion, third body interaction, formation of tribologically transform structures (TTSs) and plastically deformed areas are the main features characterizing fretting contacts [57, 63, 70, 83, 84]. The following paragraphs provide an overview of these aspects of fretting.

2.2.1 Fretting wear and debris generation – Delamination theory

As mentioned in paragraph 2.1, the fretting contact is characterized by an area of certain extension subjected to micro-movements. In this area the wear mechanism seems to be similar to the one occurring in the case of reciprocating sliding [84]. Therefore, fretting wear and the consequent fretting debris formation can be well described by the delimitation theory of wear [73, 84-88].

The surface asperities are the contact points of two mated bodies under load. Thus, if the contacting surfaces slide with respect to each other, normal and tangential loads that characterize the contact result in adhesive and plowing action exerted by the asperities. Some of the surface asperities can be deformed and fractured by the repeated (cyclical) loading action and, as a result, first debris and relative smooth surfaces are generated. The removal rate of the asperities depends on initial roughness, loading, mechanical properties of the asperities and environmental conditions [86].

The change of the surface morphology (asperities fracture) causes change in the nature of the contact between the surfaces. The asperity-to-asperity contact characteristic of the initial stage (before the sliding) is substituted by an asperity-plane contact. In each of these contacts the cyclical load, caused by the asperities plowing action, results in local plastic shear deformation. Due to the repeating loading, the plastic shear deformation accumulates, favoring cracks nucleated below the surface of contact [83, 85, 86]. At or near the surface of contact, crack nucleation is not favored due to the compressive stress existing in the contact area. Once cracks are formed, owing either to crack nucleation, void nucleation or pre-existing voids [85, 89, 90], further loading and deformation enhance subsurface cracks extension and propagation parallel to the contact surface, joining the neighboring ones [86, 90]. Cracks propagate parallel to the surface due to the compression-tension cycling loading experienced by the tip of the crack every time the asperities plow the surface [90].

At certain weak positions, cracks shear to the contact surface resulting in "delamination" of long and thin wear sheets (or flakes) forming the fretting debris [73, 86]. The thickness of the wear sheets is controlled by the depth at which cracks form and it manly depends on material properties, coefficient of friction, normal and tangential load [84-90].

If cracks cannot propagate due to a limited deformation or insufficient stress at the asperities contact (e.g. due to low friction coefficient), wear particles can still form when a sufficient number of adjacent cracks nucleates and makes the substrate weak. In this case the debris are not necessary flat and the nucleation process becomes the rate controlling mechanism [85].

Fretting debris generated as product of the delamination process are generally bigger than the ones created by the asperities removal at the initial stage of the fretting process [85]. Due to the load and the small sliding amplitude, fretting debris are likely trapped in the contact area where they can act as abrasive particles, especially if oxidized, enhancing the fretting wear. The debris can also be compacted under the contact load forming an interposing layer which might mitigate the wear process [57, 69-71].

2.2.2 Fretting fatigue

Although fretting fatigue is not the main topic of this work, fatigue and other damaging mechanisms are intrinsically related to conditions in which fretting occurs. Therefore, to understand the complexity of the fretting process, an overview of the fretting fatigue will be given.

Due to the small amount of material removed, the fretting wear is in most of the cases of small concern; however the effects deriving from such a process can be very harmful for the components lifetime. Fretting fatigue is a phenomenon arising when the surface of a component subjected to alternating bulk stresses is also fretted; this might result in severe reduction of fatigue life [57, 89-93]. For example, in Fig. 2.2 (adapted from [92]), pure fatigue and fretting fatigue for austenitic steel are compared. The fretting process reduces the fatigue resistance; as a consequence fatigue failure occurs after lower number of cycles/cyclical loads. The pits and the cracks which nucleate during the fretting process due to the sub-surface accumulation of plastic deformation (according to mechanisms explained in paragraph 2.2.1) can act as initiating points for fatigue cracks [85, 89, 91, 94-96]. Some of these cracks might not just cause formation of fretting debris, but rather propagate in the bulk, especially where the shear stresses are sufficiently high, like at the border between fretted and unfretted area. In the bulk, hundreds of μm away from the fretting surface, the influence of the contact stresses become less important; any bulk or global stresses then dominate the cracks propagation until the eventual failure of the components.

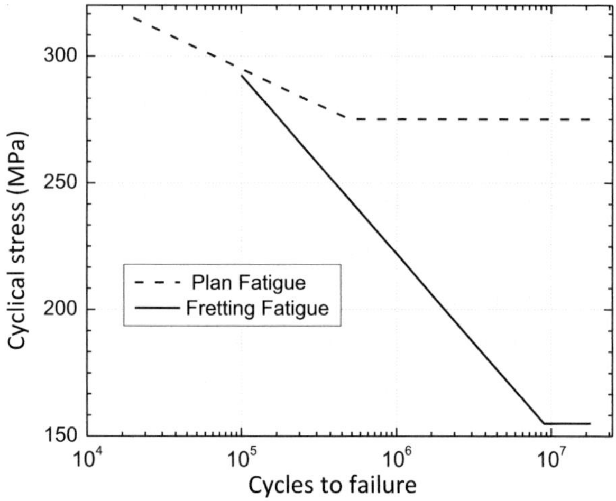

Fig.2.2. Reduction in the fatigue life due to the combined action of fatigue and fretting; graph adapted from [92].

The micro-cracks generated by fretting motion accelerate the first stage of fatigue (nucleation). Besides, according to [97] and as shown in Fig. 2.3 (adapted from [97]), the fretting fatigue cracks growth, at the initial stage, is quite faster than in pure fatigue cracking. Adhesion, oscillating movements and tensile loads between fretting asperities are main causes for the acceleration of early crack growth.

Of course, the fretting fatigue, as the fretting process in all its aspects, is susceptible to chemical, mechanical and environmental factors.

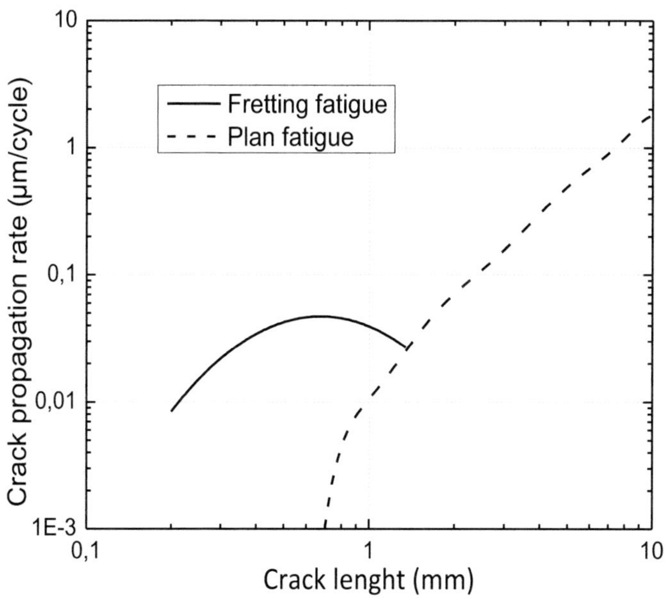

Fig.2.3. Crack propagation curve for normal and fretting fatigue; graph adapted from [97].

2.2.3 Fretting corrosion

Fretting corrosion is the result of the fretting process occurring in aggressive/corrosive environment. In such situation, as described in 1954 by H. Uhlig [98], fretting is characterized by an interplay of chemical and mechanical factors, with an observable damage, generally resulting from both. During the fretting corrosion process, the continuous mechanical action of rubbing and plowing generates clean and chemically reactive surfaces and results in formation of debris, which, together with the corrosion products, may influence the tribological behavior of the mated bodies [56, 98, 99]. Generally, most of the fretting debris consists of particles having composition resulting from the interaction environment/base material, i.e. in the case of fretting in air, debris are a mix of alloying elements of the material in contact and their oxides.

The interplay of fretting wear and corrosion mechanisms can severely affect the lifetime of components operating in aggressive environments. Indeed, for this kind of application, structural alloys characterized by the presence/development of surface protective scales, are predominately selected. Such protective layers, in the case of oxidizing environment like in most of the practical applications (e.g. air), are oxide scales. The oxide scale is formed by selective oxidation from the alloying elements of the bulk, which thereby undergoes some change in the composition. In this contest, any process affecting the stability of the protective scale can have harmful consequences [73, 99].

Due to the fretting action the initial protective oxide scale is plowed, cracked and removed by the fretting wear process, according to the delamination mechanism previously explained [73, 84, 85-90]. Oxide debris and partially metal debris are release in the contact area; the latter are quickly oxidized due to the high surface/volume ratio imparting high surface chemical reactivity. The surface emerging after the protective oxide

scale removal is characterized by high reactivity and it quickly re-oxidizes before being again abraded by the fretting action [59, 65, 73, 86, 98].
Fretting and corrosion affects each other. On one hand, oxide scales and oxide debris, acting as solid lubricant or/and abrasive particles (third body interaction), have a non negligible influence on the fretting behavior of the contact surfaces (see following paragraph 2.2.4). On the other hand, fretting action influences the oxidation process. In fact, if the fretting action removes e.g. the Cr_2O_3 scale covering stainless steels exposed to air at 650°C, nucleation and growth of Fe-Cr spinel oxide will be favored by the partial Cr depletion of the underlying material that derives from the initial Cr_2O_3 development [99, 100]. Additionally, the friction between the sliding surfaces can lead to local temperature increase, which can be significant at high sliding frequencies; this can cause the formation of oxides not expected for the test conditions and with different wear resistance [64, 101, 102].
Moreover, for some materials, the combined action of fretting and corrosion can reduce the fatigue strength. Fretting fatigue is intensified in corrosive environment like air and NaCl (e.g. sea water) compared to inert argon atmosphere [92-94, 103]. Fretting corrosion accelerates the fatigue cracking process when the protective oxide scale is removed. When this happens, the bare surface is so reactive that even in environments that are not very aggressive localized corrosion accelerates fretting fatigue. Indeed, in places where the protective scale is damaged, corrosion generates pits, which initiate cracks that might promote fatigue. The influence of corrosion mechanisms on fatigue resistance is supported by experimental results that shows an increase in fretting fatigue resistance in sea water when cathodic protection is adopted [92, 93].

2.2.4 Third body and Tribologically Transformed Structure (TTS)

In the fretting process, the central part of the contact zone is highly confined due to the small sliding displacement. Fretting debris formed by plastic deformation (abrasion, adhesion, fatigue etc.) and by fracture of the superficial layers of the mated bodies (first bodies) are trapped in the contact zone for a certain period before being released [71, 104]. This situation favors the occurrence of the so called third body interaction and the formation of the tribologically transformed structure (TTS), shown in Fig. 2.4 (adapted from [83]) and generally not observable in other tribological systems.

When fretting occurs in oxidizing environments like air, the oxide debris trapped in the contact area can either act as abrasive particles increasing the wear rate or can be compacted in a sort of protecting film that slows the wear process. Thus, in fretting processes, after few cycles the situation changes from a two to a three bodies contact. This change must be considered for the interpretation of the experimental results.

The third body consists in fretting debris which are trapped, broken, oxidized, deformed and aggregate on the contact surfaces (Fig. 2.5 adapted from [105]). After the formation of the third body, the wear process will keep evolving; therefore the presence of the third body might affect the whole wear system. Besides, in the third body the particles can change size, morphology and composition with the evolution of the fretting process (Fig. 2.5) [70, 71].

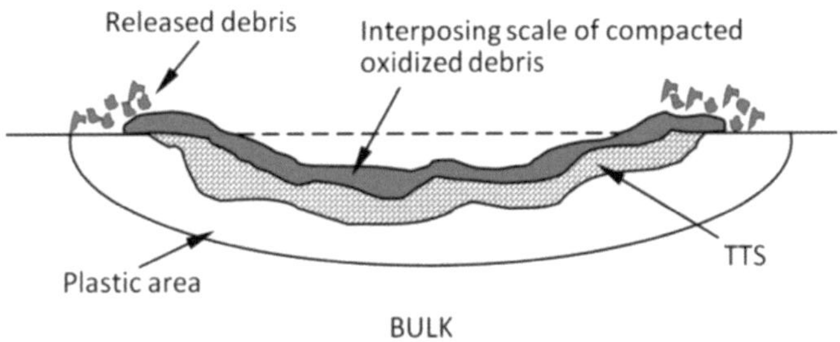

Fig.2.4. Scheme of the cross-section of a fretted surface. Oxide bed, detached debris, TTS and plastically deformed area can be recognized.

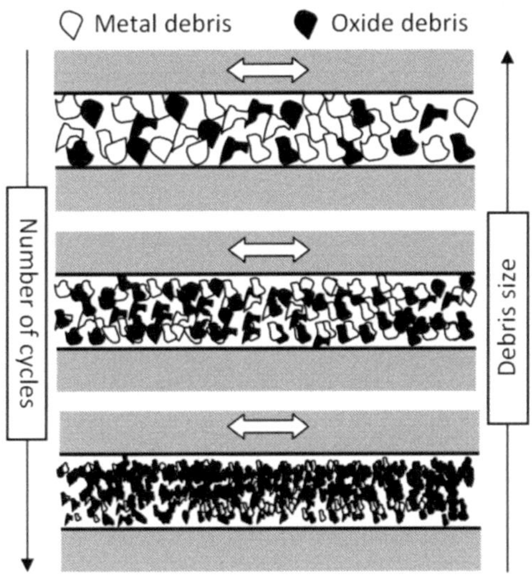

Fig.2.5. Generation and evolution of the fretting debris (third body) during the fretting process; figure adapted from [105].

The role of the third body during the fretting process was matter of investigation in different researches. In fretting of mild steel in air, when the sliding amplitude is low enough to favor the debris retention in the fretting area, the fretting wear becomes self-limiting because of the formation of an interposing scale made of compacted oxidized debris. Such scale supporting the contact load and keeping separated the sliding surfaces prevents the metal-to-metal contact [88, 91, 106-114]. The third body, as interposing oxide bed, acts as a sort of lubricant decreasing friction coefficient and wear rate [70, 71].

Conversely, it was also noticed that free debris and particles removed from the oxide bed might exert an abrading action, increasing the fretting wear rate, as long as they are not released from the contact area. Therefore, to determine the wear behavior of the later fretting process the rheology of the third body is also important.

The time necessary for the formation of the third body generally depends on materials and environmental factors. In some cases, the compacted oxides bed formation was noticed already after 10000 fretting cycles [71].

The Tribologically Transformed Structure (TTS) is the region from where fretting debris are generated and is located between the plastically deformed area and the compacted debris (Fig. 2.4). The TTS is characterized by the same composition as the initial bulk material but with a nanocristalline structure [69, 83, 105, 115]. This supports the hypothesis that TTS is neither an aggregate of fretting debris nor a mixture of the materials brought into contact. Additionally, different from fretting debris, in TTS low oxygen content is detected (comparable to the one of the bulk material). TTS is generally very brittle and crossed by many cracks. TTS is characterized by high hardness, 2-3 times higher than the original bulk material but Young´s modulus seems to remain the same as for the bulk [69, 83, 105].

A well established TTS layer is generally formed after little less than 1000 cycles. However, at specific test conditions already after few tens of fretting cycles TTS was observed. Besides, after a large number of cycles ($>10^6$), TTS seems to disappear; this is most probably the reason why generally it is not mentioned in industrial failures analysis [83, 105].
Different hypotheses have been proposed to explain the formation mechanism of TTS and questions about it are presently still open.
The flash-temperature created by the friction between sliding surfaces is indicated as possible cause of the TTS formation [115]. The authors suggest that in a two step process, first a local temperature increase and second quenching due to the substrate acting as a heat-sink, a martensitic structural change occurs generating the TTS. The weak point of this theory is that TTS was noticed also in fretting cases at very low frequencies (< 1 Hz) when the temperature increase hardly exceeds 100°C.
After several observations and analyses, it was supposed that TTS was the result of high plastic strains together with transfer phenomena [116-118]; such structure was named "mechanically mixed layer" in reference to the mechanical alloying [119]. According to this model, a second transferred phase is necessary to generate very fine grains and the transformation occurs only if both chemical and mechanical processes are present. When in the fretting contact the counterpart cannot provide the second phase (i.e. in homogeneous contacts), the environment (e.g. oxygen) may enable the mentioned chemical process. However, the role played by oxygen remains uncertain and no material transfer was ever detected in the TTS, even in heterogeneous contacts.
A more recent and complete hypothesis for TTS formation starts from the assumption that dislocations generated by inclusions are responsible for recrystallization in presence of a stress field [83, 105, 118, 120]. Under the fretting contact load, the difference in elastic properties (mainly Young´s modulus) between matrix and inclusions (i.e. minor alloying elements)

induces high plastic strains, which lead to re-crystallization of the microstructure through a nucleation/growth process starting from the surface, as schematically shown in Fig. 2.6 (adapted from [83]). Nano-grains formed during the re-crystallization process act themselves as stress concentration sites so that new nuclei of TTS are created. Although some kind of alloying elements re-distribution or segregation may occur in the nano-scaled crystal grains, at the composition analysis scale of ~1 μm (limit value for most of the characterization methods) the TTS would result with the same composition as the base material.

According to an energy approach, the TTS formation can be related to a certain critical dissipated energy. There is a threshold energy, corresponding to a certain plastic work density, below which the microstructure is unchanged and above which TTS nucleates and grows. From this approach it also appears that increase in amplitude, number of fretting cycles and normal load favor TTS formation more because more energy is dissipated [105].

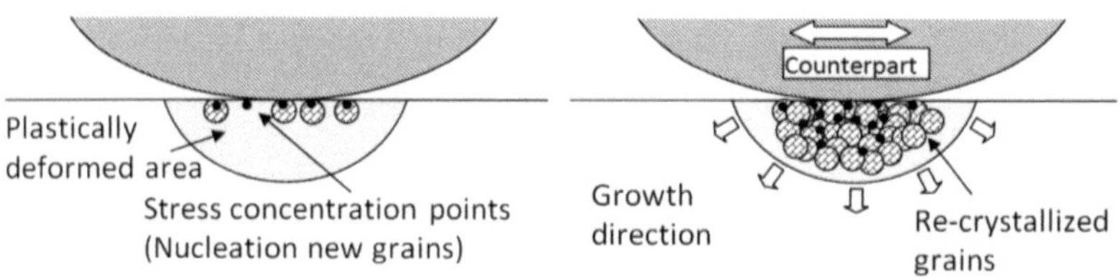

Fig.2.6. Nucleation-growth model and re-crystallization of the TTS due to concentrated high strain. Figure adapted from [83].

2.3 Parameters affecting fretting

Up to 50 parameters affecting the fretting process can be possibly identified [72].They can be divided in three main groups: mechanical variables, such as applied load, sliding amplitude, friction force and frequency; material properties such as relative hardness, material composition and roughness; and environmental variables such as temperature, oxygen content, presence of corrosive substances, relative humidity and presence of lubricants. Hereafter, the influence of some of these variables is shortly described.

2.3.1 Number of cycles

In most quantitative measurements of fretting on steel, the volume of material removed by the fretting action, after a certain run-in period, increases nearly linearly with the number of cycles [1, 66-68]. The run-in period can differ in extension and it is characterized by lower or higher wear rates depending on the others fretting variables (e.g. environment and contact paramenters). The existence of a run-in period is probably related to the fact that some typical features of fretting, such as TTS formation and third body interaction, develop in the first thousands of fretting cycles and influence the fretting process evolution. As shown in Fig. 2.7 (adapted from [121]), for steels tested in air, generally the run-in period presents a higher wear rate due to the initial fast cracking and removal of the covering oxide scale, resulting in fretting debris formation.

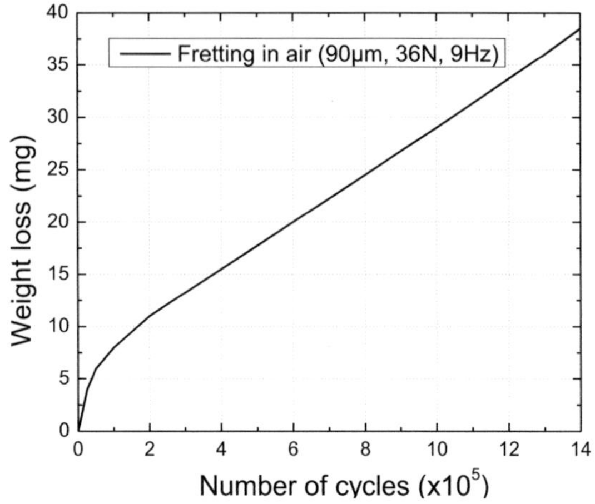

Fig.2.7. Fretting wear (weight loss) of mild steel in air versus number of cycles. Fretting test parameters: 36 N, 90 µm, 9 Hz. Plot adapted from [121].

2.3.2 Applied load

The fretting contact description given in paragraph 2.1 and the outcomes of different experimental activities (e.g. [87]), suggest that variations in the applied load influence the contact conditions, e.g. the threshold tangential force to induce displacement and extension of the micro-sliding area in the fretting contact.

The fretting damage in terms of wear (i.e. weight and/or volume loss) seems to be affected by the applied load in relationship with the imposed sliding amplitude. Results of fretting test performed on different Ni-based alloys (i.e. Inconel 600 and Inconel 690), varying applied load and imposed sliding amplitude, show that the fretting wear increases nearly linearly with the load up to a certain turning point, after which it decreases (Fig. 2.8 adapted from [124]). The value (position in the graph) of such turning point and the trend of the plot around it are influenced by several variables.

In this sense, sliding amplitude and type of material play a key role [69], for a longer imposed sliding amplitude higher values of the turning point were measured.
Additionally, the contact load influences the fatigue resistance; generally higher loads shorten the fatigue failure time [70].

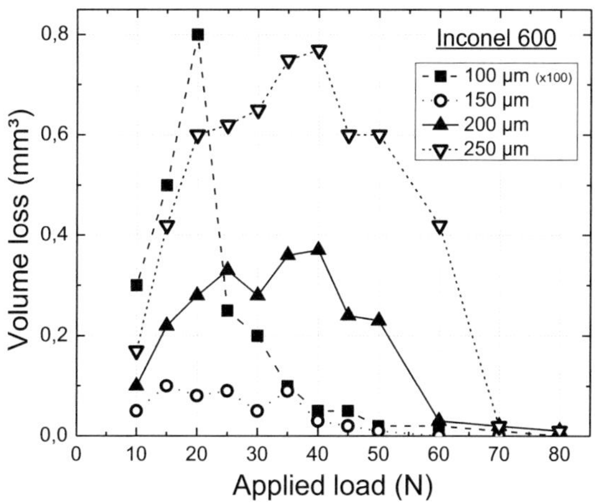

Fig. 2.8. Fretting wear (volume loss) of Inconel 600 versus applied load, for different imposed sliding amplitude. Volume losses of the 100 µm fretting tests are reported multiplied by a factor 100. Plot adapted from [124].

2.3.3 Amplitude of the slip

The sliding amplitude determines the length of the fretting wear path; therefore it is quite intuitive that more significant material loss corresponds to longer amplitude of the slip. From the earlier investigation of fretting of steels in air, a linear dependency fretting wear- amplitude was noticed [121]. However, later works showed that the fretting wear damage increases linearly with amplitude only above a certain threshold, corresponding to amplitudes about 100 µm (Fig. 2.9 adapted from [126]). For smaller values, the amplitude effect must be considered in relation with the others parameters (i.e. load and environment variables) [65, 126, 127].

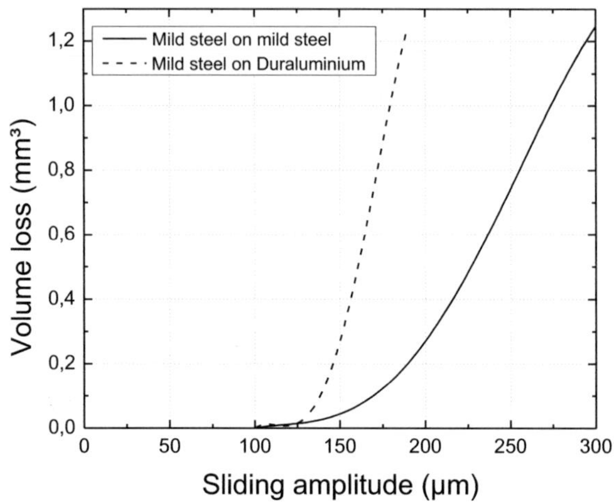

Fig.2.9. Fretting wear (volume loss) vs sliding amplitude. Data relative to fretting contacts: mild steel-mild steel and mild steel-duraluminium in air, for 10^6 cycles and with 19 N applied load. Plot adapted from [65].

Moreover, bigger amplitudes favor fretting debris release from the contact area, which on one hand shorten the abrasive action of the oxidized particles, on the other hand hinder the interposing oxide bed formation.

2.3.4 Frequency

Frequency can affect the fretting process under different aspects. There are some experimental indications that at lower frequencies the fretting damage is more evident (Fig. 2.10). Indeed, low frequencies favor the development of thicker oxide scale before it is removed by the fretting action.

This might result in more significant amount of material worn away by fretting at every cycle [98, 102, 104, 128, 129].

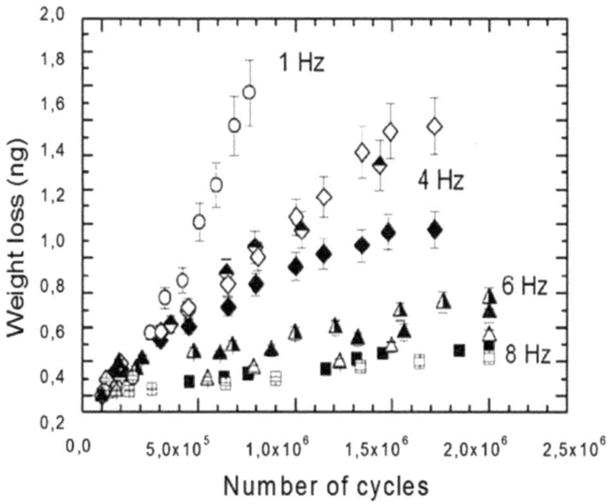

Fig.2.10. Fretting wear (volume loss) as function of frequency. Fretting tests performed on Ti-6Al-4V alloy: 1-4-6-8 Hz, 40 µm, 60N, for ~10^6 cycles. Plot adapted from [129].

Furthermore, high frequencies can lead to non-negligible local temperature increase, which may lead to re-crystallization [64] and influences the oxidation process. Besides, the frequency variations influence the mechanical properties of the material through temperature and strain rate hardening effect [71]. Frequency might affect the fretting fatigue properties. For imposed values of the sliding amplitude, there are frequency ranges in which the fatigue behavior is more favorable than in others [62].

2.3.5 Temperature

Temperature is surely one of the most important physical variables affecting the fretting process. It influences both mechanical properties of the material (e.g. ductile-brittle behavior) and corrosion kinetics. As a result, fretting wear, fretting corrosion and fretting fatigue might significantly change according to the operating temperature range. Considering for example fretting of steels in normal atmosphere, the results of several experimental activities support the existence of transition temperatures at which the fretting behavior changes [99, 130-133]. With reference to Fig. 2.11 (adapted from [130, 131]), the fretting damage decreases up to 200°C (first transition temperature) where it remains constant up to about 300°C and then it decreases again up to 500°C (second transition temperature). For higher temperatures a faster increase of the fretting damage was noticed. The variation with the temperature of the fretting behavior is explained considering oxidation kinetics and the type of oxide forming at different temperatures. Around 200°C the oxidation rate changes from logarithmic to parabolic [130]. The thicker and more adherent the oxide scale the higher the wear resistance and the ability to act as solid lubricant [130-134]. Moreover, different iron oxide scales characterized by different fretting resistance, develop at different temperatures. Magnetite (Fe_3O_4), which is the prevailing oxide above 380°C, has higher fretting resistance than Fe_2O_3 that is the main oxide up to 200°C and which behaves better than FeO developing above 500°C [130-137]. In addition, at temperature above 500°C the thickness of the oxide scale can be excessive for being stable and it is removed under fretting action so that more significant fretting damage occurs.

Similar results were reported for titanium alloys [138] and nickel alloys [82, 139].

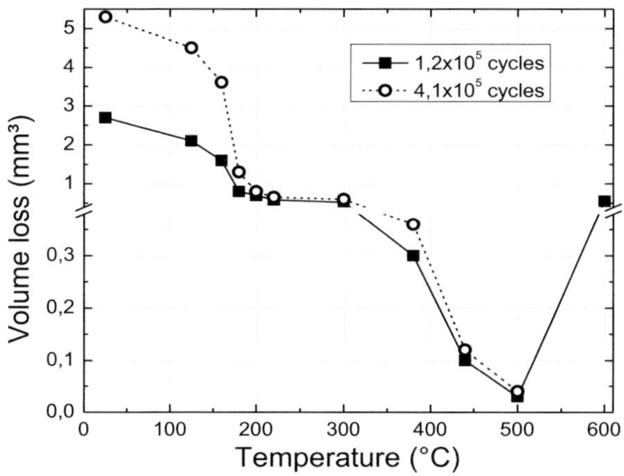

Fig.2.11. Fretting wear (volume loss) as function of temperature. Fretting tests performed on mild steel: ~7 Hz, ~15 N, 75 μm and up to $4.1 \cdot 10^5$ cycles. Plot adapted from [130, 131].

2.3.6 Environmental variables

Atmospheric oxygen and water have a strong effect on the fretting process, especially for metals [56]. It is widely recognized that oxygen and water accentuate fretting wear and surface damage while the inert atmosphere (e.g. argon or nitrogen) mitigate fretting of metals [121, 140]. Fretting wear in nitrogen is around 3 times lower than in air [56]. The role of the atmospheric pressure is also very important. In [60] it is reported that from 10^{-3} to 10^{-1} Pa the friction coefficient decreases by about a factor 2.

The presence of lubricants favors reduction in fretting wear. In general, lubricants reduce the friction between the sliding surfaces (Fig. 2.12 adapted from [141]), the adhesion between the asperities and prevent the corrosion wear, so that the fretting wear proceeds by milder fatigue-based delamination wear mechanisms [141-147].

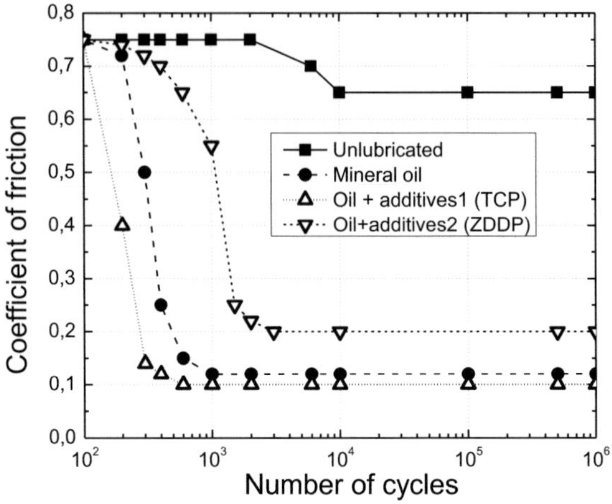

Fig.2.12. Coefficient of friction versus number of cycles for unlubricated and lubricated fretting contacts. Fretting tests performed on SUJ2 alloy: 6 Hz, 49 N, 90 μm, up to 10^6 cycles. Plot adapted from [141].

2.3.7 Material features (hardness, roughness and surface coatings)

The role of hardness is not totally clear so far. Hardness influences fretting by affecting the delamination process, the friction coefficient and the abrasive power of the debris [93]. However, from more recent research activities (e.g. [143, 149]) it emerges that hardness has no direct relation with fretting wear and its manipulation was found to be an unreliable means for improving fretting resistance.

Clearer is the role of the microstructure, which influences the fretting wear rate. Martensitic structure shows better wear properties and fretting fatigue resistance, especially at high cyclic stresses, than the austenitic structure [150, 151].

It is commonly recognized that a high degree of surface finishing, maximizing the contact points, increases the fretting damage; thus, rough surfaces are preferred to minimize the fretting effect [152, 153]. At high temperature the converse is true; surface with better finish suffer less of fretting than rough surfaces [154].

Additionally, many scientific activities pointed out the positive effect of surface coatings with respect to the fretting resistance [147, 155, 156]. Different methods and types of coatings with the related wear resistance properties have been investigated. There are some indications that surface hardened alloyed layers obtained by laser alloying have a beneficial impact on fretting properties, while thin coatings deposed by PVD showed no significant influence on the fretting wear process [147].

2.4 Fretting maps

In the previous paragraphs, the main mechanisms (e.g. corrosion, fatigue and wear) involved in fretting have been described, together with an overview of the principal fretting characteristics and affecting parameters. Since the occurrence and the extension of different types of surface degradation are related to fretting conditions and parameters, the concept of fretting maps [62], which identify the fretting regimes, will be hereafter explained.

In practical applications it is important to determine the fretting regime resulting from a given set of parameters. A powerful tool for this purpose is the concept of fretting maps, which were firstly introduced by Vingsbo in 1988 [62] in order to compact in graphs mechanical, physical and environmental variables affecting fretting.

It can be rather difficult to foresee in which fashion the fretting mechanisms interacts with each other and when and how extensively a certain mechanism occurs or prevails over the others. For example, as reported in the description of the third body interaction, in certain conditions, formation of compacted oxide bed mitigating the wear rate is favored.

As detailed in paragraph 2.1, experimental observations and theoretical analysis showed that in fretting the contact conditions change with displacement amplitude and applied load. On the other hand, the contact conditions influence the fretting mechanisms.

A fretting map is a diagram showing the relevant fretting regimes in two variables, e.g. load-displacement amplitude, and in which plotted lines represent the critical values for the transition from one fretting regime to another (regime boundaries). Each regime is characterized by certain contact conditions, fretting mechanism of surface degradation and resulting fretted surface appearance [62, 157]. Thus, fretting maps can also be an alternative approach for classifying, representing and interpreting experimental data. Besides, a successfully and correctly recorded fretting map of a certain tribological system can facilitate the choice of operating conditions adequate to alleviate fretting effects.

In [62], fretting maps relative to different experimental conditions are built. Examples of fretting maps as function of load - amplitude, load - tangential force, specific wear coefficient/fatigue life - amplitude, specific wear coefficient/fatigue life - load and frequency - amplitude, are reported in Fig. 2.13, 2.14, 2.15, 2.16 and 2.17 respectively (adapted from [62, 157]).

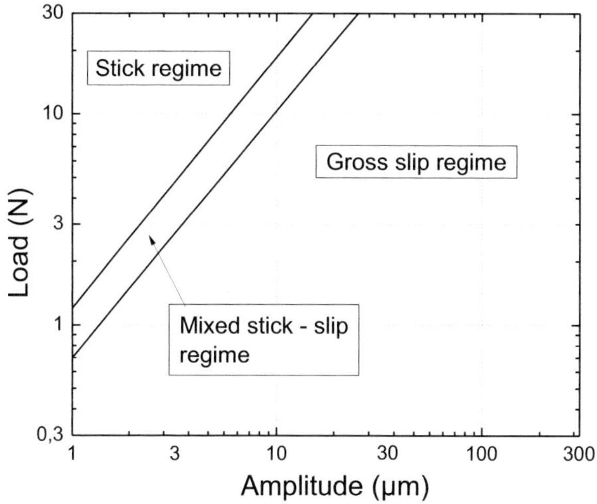

Fig.2.13. Fretting map of applied load versus amplitude. Graph adapted from [62].

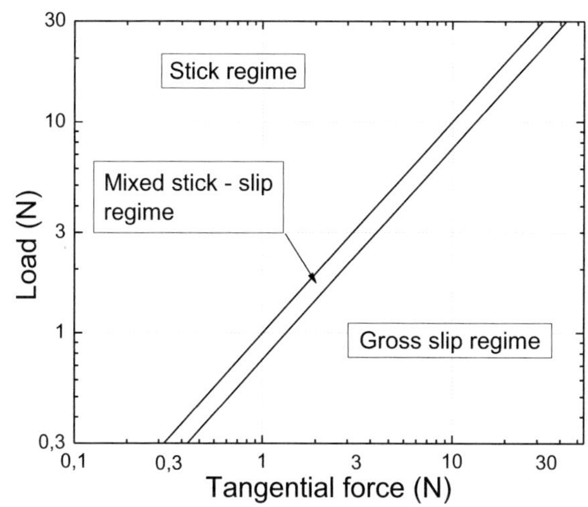

Fig.2.14. Fretting map reporting applied load versus. tangential force for constant sliding amplitude. Graph adapted from [62]

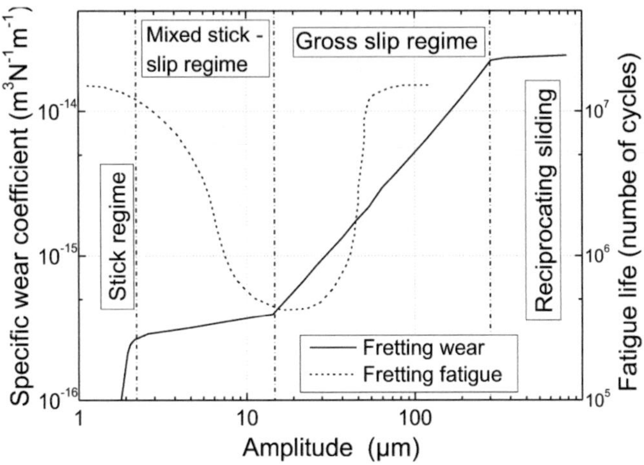

Fig.2.15. Fretting map reporting fretting wear and fretting fatigue versus sliding amplitude at constant load. Graph adapted from [62].

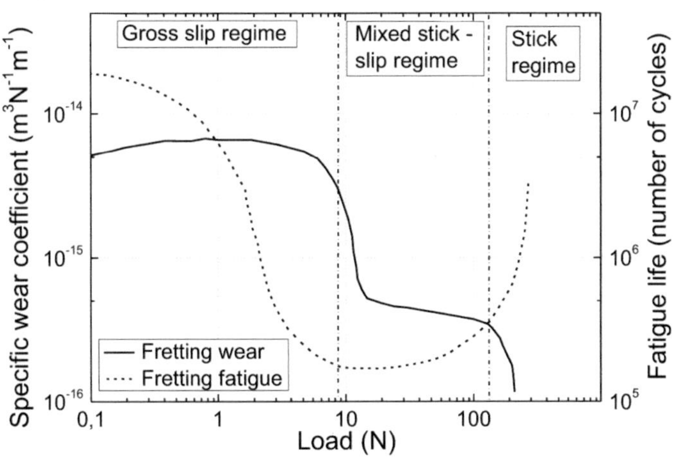

Fig.2.16. Illustration of a fretting map reporting fretting wear and fretting fatigue life vs load for constant sliding amplitude. Graph adapted from [62].

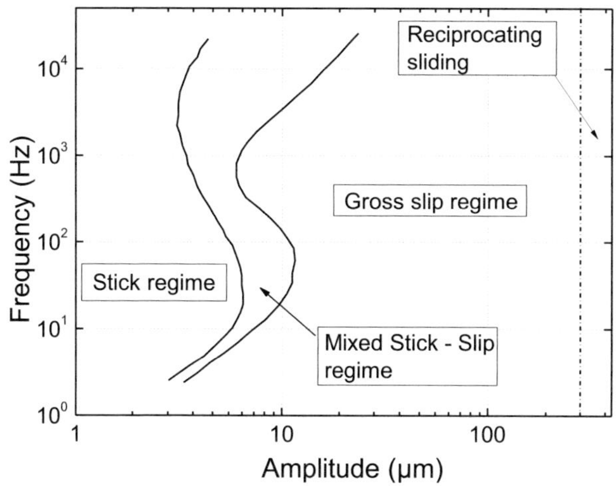

Fig.2.17. Fretting map reporting the sliding frequency versus the sliding amplitude with constant applied load. Graph adapted from [62].

In a fretting map up to 4 regimes can be identified:

1. **Stick regime**. It characterizes fretting contact with very small displacement amplitudes which leads to very limited fretting surface damage by wear and corrosion; besides, generally no fatigue cracks are noticed for a low number of cycles ($<10^6$). The surface morphology indicates contact by multiple asperities but the material between the contact points remains mostly unaffected (Fig. 2.18), suggesting that elastic deformation in the near-surface regions of the two components accommodates the sliding between the loaded contacting surface. The adhesive joining between the asperities, which can be plastically sheared in the fretting direction (scratches in Fig. 2.18), maintains the interface under stick condition.
This fretting condition can be defined as a low damage fretting.

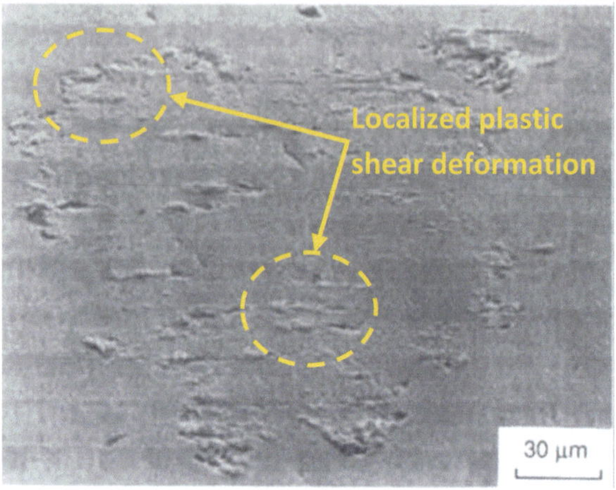

Fig.2.18. Fretting wear scar characteristic of stick contact conditions. Fretting test performed on Niobium: 100 Hz, 11 N, 2 µm, 10^6 cycles [62].

2. **Mixed stick-slip regime (or partial slip regime).** Like for the stick regime, there is a stick region (central part Fig. 2.19), where the surface damage is restricted to plastic shear of asperities, giving only a limited contribution to the total surface damage. In the annular slip region extensive crack formation occurs, especially at the border to the stick region where the highest stresses are generated, higher than the one generated by an individual asperity contact. Fretting debris are produced by crack propagation and intersection, and may act as abrasive particles, especially if oxidized, until they are eventually released from the contact area. Generally, wear and corrosion effects are small but accelerated crack growth, starting from fretting cracks, may result in significantly reduced fatigue life. In this fretting regime, the predominant mechanism is fretting fatigue or corrosion enhanced fretting fatigue (see paragraph 2.2.3) [62, 63, 87].

Fig.2.19 Fretting wear scar characteristic of mixed stick/slip regime (partial slip). Fretting test performed on AISI 304 with 100 Hz, ~ 11 N, 4 µm, 10^5 cycles [87].

3. **Gross slip regime.** In this condition, all the asperities are broken in each cycle. The morphological analysis of the fretted surface cannot distinguish the single asperity contact (like in Fig. 2.18) because every asperity in the fretting process slides across several others on the opposite surface. The fretting damage consists in sliding wear grooves in the fretting direction and the extensive plastic deformation results in a structure of scales which will eventually delaminate (Fig. 2.20). In oxidizing environment, cracking and delamination will be enhanced, due to the brittle nature of the oxides. The fretting debris are transported towards the contact rim and crushed in small fragments. In this regime, wear and corrosion (e.g. oxidation) are strongly coupled and predominant.

4. **Reciprocating sliding regime**. Sliding wear occurs in this regime.

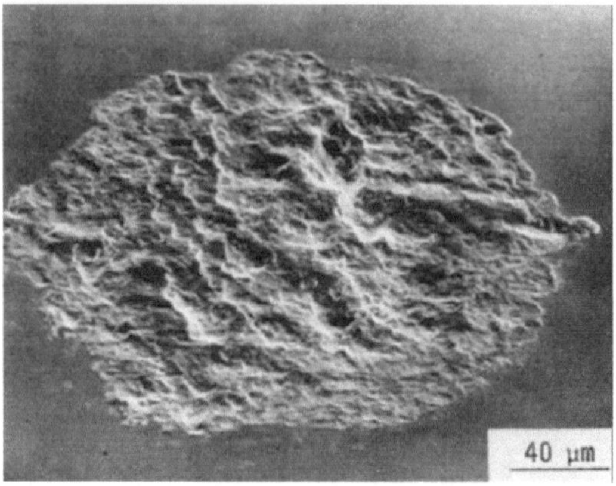

Fig.2.20. Fretting wear scar characteristic of gross-slip regime. Fretting test performed on AISI 304: 100 Hz, ~ 11 N, 16 µm, 10^5 cycles [87].

In fretting maps, variations of the main affecting variables (e.g. environmental or contact conditions) can significantly change the position of regime boundary lines [62, 157]. Accordingly, the presence of lubricants for instance, as shown in Fig. 2.21 (adapted from [88]),or the development of an interposing bed of compacted oxides acting as a solid lubricant, lower the friction between sliding bodies and cause shifts in the regime boundary lines of the fretting maps [158].

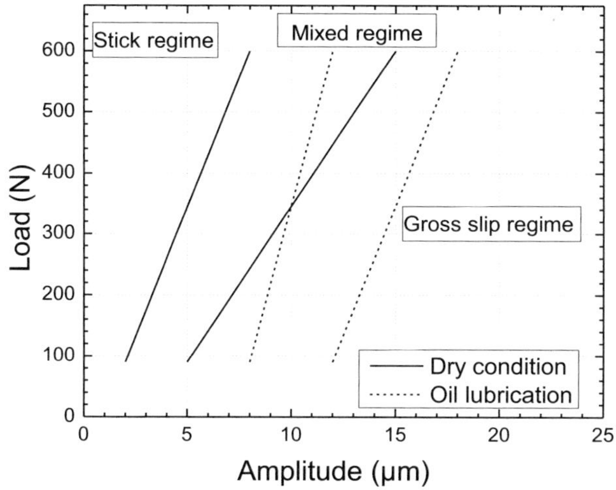

Fig.2.21. Fretting map load - amplitude for AISI 1045/AISI 52100 steel for 10^4 cycles at 1 Hz in dry and lubricated conditions [158].

2.5 Fretting in nuclear power plants (NPP)

Fretting wear, fretting corrosion and fretting fatigue have great impact for safety and economics of the nuclear industry [159]. Starting from the 1970s, several authors carried out empirical and analytical investigations aimed to characterize fretting occurring in light water (e.g. PWR) and heavy water nuclear reactors (e.g. CANDU) [60, 61, 124, 159-167].

Due to the so called flow induced vibrations (FIV) , the components most susceptible to fretting damage are steam generator (SG), heat exchanger (HX) tubes, fuel channels, fuel cladding (FC) and pressure tube (PT) [159]

As described in chapter 1, in nuclear power plants, the coolant is aimed to remove the fission reaction heat from the reactor core and to carry it to the SG where the thermal cycle for electric energy production starts. From the point of view of the coolant, the aforementioned reactor components are essentially cylindrical structures (piping, fuel channels) or bundles of cylinders (fuel bundles, tube bundles) subjected to axial or cross-flow, internal or external to the cylinders and sometimes confined in narrow annuli or in close-packed bundles. These cylindrical structures are often supported in several places, e.g. heat exchanger tubes are supported by baffle-plates, fuel elements by bearing pads and support/spacer grids, and piping systems by piping supports [168].

In this context, particular flowing conditions might generate FIV causing reciprocating movements, under load and with certain amplitude, of reactor components against corresponding supports. Thus, favorable conditions for fretting might occur [124, 169].

The FIVs are dynamic forces generated in nuclear reactor components by fluid flow and causing them to vibrate. Four FIV excitation mechanisms are relevant for nuclear reactor components: (i) fluidoelastic instability, (ii) periodic shedding, (iii) turbulence induced excitation and (iv) acoustic resonance [168]. Generally appropriate design solutions are adopted in order to avoid the conditions for FIV mechanisms establishment. However, during the component operating lifetime, dimensional instability or partial damaging due to the reactor extreme environment, i.e. irradiation damage and high thermal-mechanical loads, might favor the advent of FIV [168-171].

The fretting related problems are addressed from different perspectives. Designers and manufactures are concerned to create reliable models to predict the fluid-structure interactions and the tribological response of some critical structural elements to FIV. Operators of NPP need data about fretting in operating conditions to optimize the critical components management for attaining the maximum economic benefit and service life without sacrificing integrity and safety [159].

Fretting in advanced lead alloy cooled nuclear systems

According to the designers' predictions and models, in next generation lead-alloys cooled nuclear systems (e.g. LFR), the area expected to be most affected by fretting are the HX tubes in the steam generator and the fuel claddings in the reactor core (Fig. 2.22).

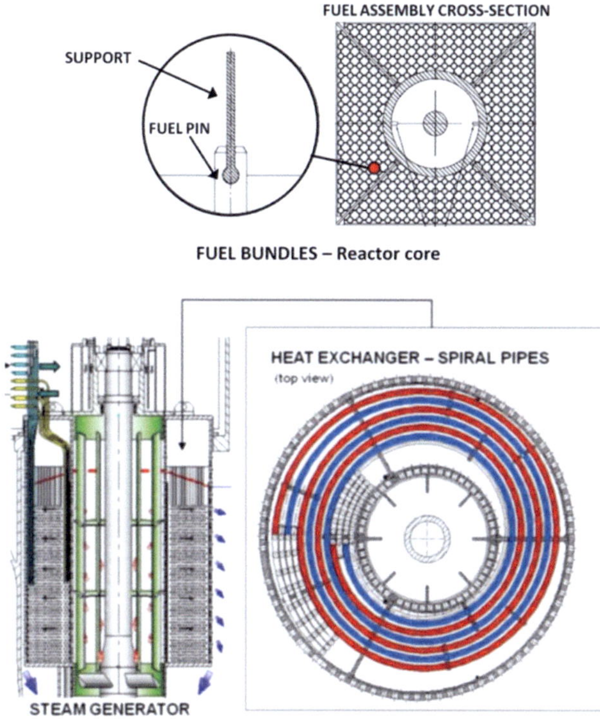

Fig. 2.22. Illustration of areas expected to be affected by fretting in a lead alloys cooled nuclear system. In the upper image, a typical fuel assembly with the particular of the connection fuel pin-support is visible. In lower image, the design of ELSY SG with the peculiar spiral HX can be seen.

For these components, fretting displacement with frequency up to 20 Hz and amplitude up to 200 μm are expected [172]. In the particular case of a

LFR or ADS, fretting process is occurring in an environment characterized by the presence of liquid lead/lead-alloy with a certain amount of oxygen in solution (typically 10^{-6}-10^{-8} wt%) and temperature in the range 300-600°C. Thus, interaction between fretting, corrosion mechanisms affecting structural steels in lead-alloys (i.e. oxidation and dissolution) and corrosion barriers (e.g. protective oxide scale and protective coatings) might occur, jeopardizing the components performance and lifetime.

3. Experimental equipment for fretting test in heavy liquid metals

3.0 Introduction

To investigate extension, modality and characteristics of fretting and its related phenomena (see Chapter 2) in liquid Pb/Pb alloy cooled nuclear systems, a particular experimental setup is necessary. At this scope, a test rig capable to perform fretting tests in heavy liquid metals (HLMs) in reactor relevant conditions (see chapter 1) should be deployed. The constant monitoring of fretting variables (e.g. amplitude, load, frequency and friction force) as well as principal environmental parameters (e.g. temperature and oxygen content in HLM) is a further essential feature for such test rig. However, such device is not available on the market. As consequence, a significant part of this work was to design and realize a dedicated facility aimed to simulate fretting in possible nuclear reactors friction contacts.

Although a lot of research activities were carried out to characterize fretting wear for different operating conditions and materials (e.g. [56-66, 74-89, 96-108, 130-150)], fretting corrosion in liquid Pb/Pb alloy has not been addressed properly so far. Starting from the designs of fretting rigs for test in air and aqueous environment [102, 173-177], and applying some of the solutions adopted in facilities used for corrosion tests in HLMs [178], a novel facility was assembled.

This facility, named FRETHME (FREtting corrosion Test in Heavy liquid MEtals) and already detailed by the author of this work in [179], simulates fretting through the contact between oscillating specimens and counter specimens under load. The main affecting parameters like amplitude, frequency, applied load and friction force are controlled and monitored during the tests by appropriate sensors. Temperature and oxygen content of the Pb/Pb alloy are determined by suitable heating and gas supply systems.

3.1 Design and operating requirements

To simulate the fretting process taking place in the reactor core and SG of a Pb/Pb alloy cooled nuclear system, the test facility must generate oscillation frequencies in the range 1-40 Hz and impose very small and controllable slip amplitudes in the 10-200 μm range. Furthermore, operating periods up to thousands of hours (10^6-10^8 cycles), which is the time range of interest for investigating fretting wear and fretting corrosion in molten Pb/Pb alloys, are required. In addition, to match conditions relevant for LFRs and ADSs, the fretting corrosion facility has to operate between 350 and 650 °C and with controlled content of oxygen in the liquid Pb alloy, typically in the range 10^{-6}-10^{-8} wt% [4, 6, 23, 180].
For studying the fretting phenomena and for interpreting test results, it is essential to identify and consider the main affecting parameters. Challenging for the realization of the fretting corrosion facility was to adopt adequate design solutions and to choose on line sensors to control and monitor amplitude of the slip, load, friction force, frequency, oxygen content and temperature.

3.2 The FRETHME facility

The 1.4571 austenitic steel was selected as construction material for the FRETHME facility. Affordable cost, high workability and weldability, good mechanical properties at high temperature and good resistance in aggressive conditions are all characteristics that have favoured the choice of this alloy.
The FRETHME facility consists of two heated chambers: a conditioning and a testing chamber. In the conditioning chamber (1, Fig. 3.1), the Pb/Pb alloy is melted and conditioned in terms of temperature and oxygen content. In the testing chamber (3, Fig. 3.1) the fretting process is simulated by loading fixed counter-specimens on sliding specimens (3-4, Fig. 3.2).

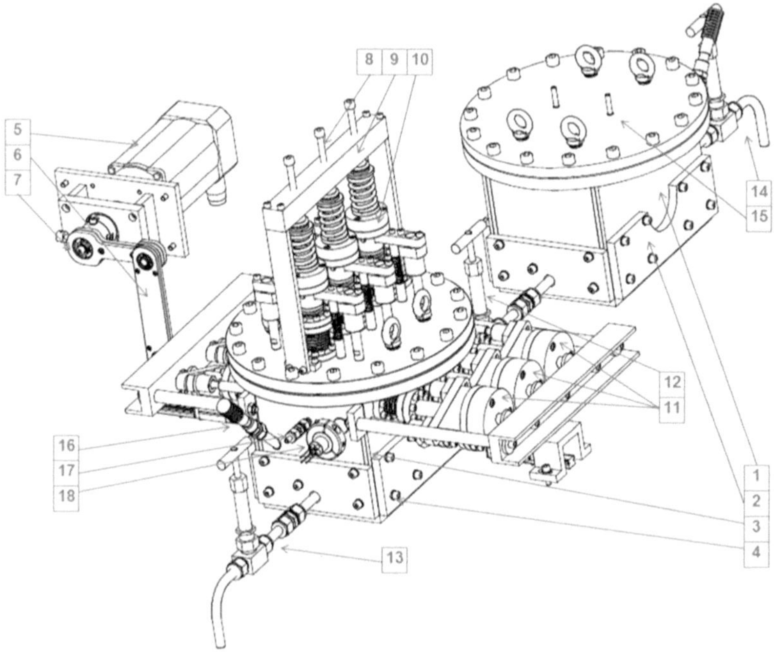

1. Conditioning box, 2. Heating system conditioning box, 3. Testing box, 4. Heating system testing box, 5 Servo electric engine, 6. Crank assembly, 7. Eccentric clutch, 8. Screw for applying the load, 9. Sprung load cells group, 10. Strain-gauge sensor for load, 11. Strain gauge sensors for tangential force, 12. Connecting channel valve, 13. Emptying valve testing box, 14. Emptying valve conditioning box, 15. Gas supply channels, 16. Oxygen sensor, 17. Thermocouple, 18. Level sensor

Fig.3.1. Overview of the FRETHME facility assembly

Testing and conditioning chamber are connected by a heated channel, which is opened/closed by a valve (12, Fig. 3.1).

The facility is designed to test three couples specimen/counter-specimen at the same time (2, Fig. 3.2). The shape of the specimens can be either cylindrical or prismatic, while the counter-specimens are flat. Dimensional details of specimens and counter-specimen are given in the next chapter.

In the testing chamber, the specimens are placed on a metallic carter (5, Fig. 3.2) connected to a crank assembly driven by a servo-electric engine

(5-6, Fig. 3.1). An eccentric clutch (7, Fig. 3.1) converts the rotatory motion of the drive motor (max frequency = 3000 rpm = 50 Hz) into linear reciprocating motion. Controlling the eccentricity of the clutch, the oscillation amplitude of the carter can be set in the range 10-1000 μm.

The counter-specimens are fixed on rigid vertical arms (1+3, Fig. 3.2) connected to sprung load cells (9, Fig. 3.1; 6, Fig.2). The contact between specimens and counter-specimens is achieved by lowering the mentioned rigid vertical arms via turning displacement screws (8, Fig. 3.1; 7, Fig. 3.2). Contact loads in the range 0-1000 N can be applied, corresponding to a contact pressure in the range 0-500 MPa (for the contact geometry).

To melt and maintain the Pb/Pb alloy at the desired temperature, a heating system, which consists of metals plates (Maier Elektrotechnik ®) placed on the walls of the facility chambers, is adopted (2+4, Fig. 3.1; 13, Fig. 3.2). In these plates, the heating is produced by resistance wires, which allow a maximum working temperature of 700 °C. The channel connecting the two chambers is wrapped and heated by resistance wires (Maier Elektrotechnik ®) that can operate in the same temperature range of the FRETHME facility chambers.

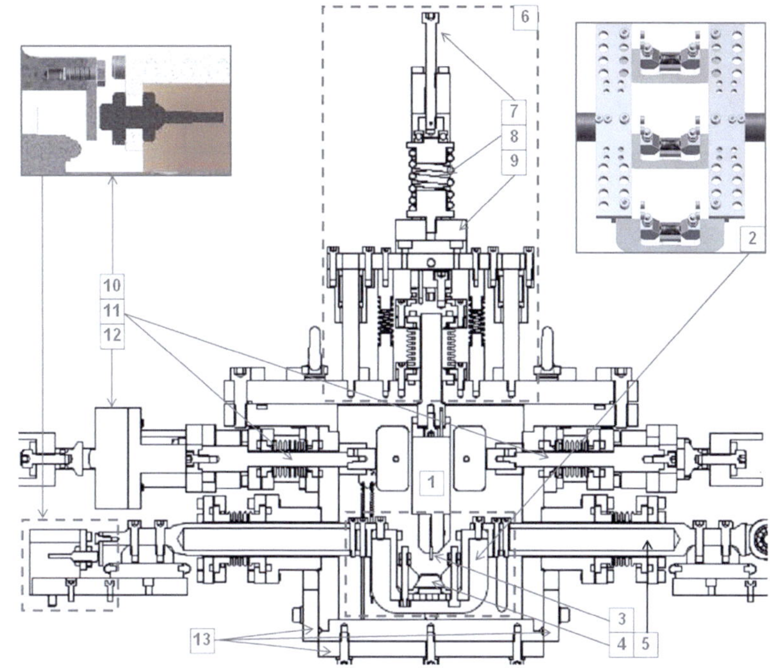

1. Counter specimen holder; 2. Specimens holder; 3. Counter specimen; 4. Specimen; 5. Oscillating carter; 6. Load cell; 7.Screw for applying load; 8. Load cell spring; 9. Load strain-gauge sensor; 10. Eddy current sensor; 11. Pressing rods tangential force apparatus; 12. Tangential force strain-gauge sensor; 13. Heating plates

Fig.3.2. Cross-section of the FRETHME facility testing chamber.

3.3 Sensors and data acquisition

Monitoring the mechanical contact response and the environmental conditions during the experimental tests is mandatory for a successful assessment of the fretting corrosion damage; this was a key issue for the facility realization. Appropriate sensors and design solutions have to be adopted to observe the main affecting parameters of fretting corrosion in molten Pb/Pb alloys.

The position of the counter-specimens, which are vertically movable in the molten Pb alloy, has to be known for the correct execution of the fretting test. Potentiometric displacement transducers BURSTER 8712 are used to address this issue. The position of the counter-specimens can be determined with a sensitivity of 10 μm.

In the FRETHME facility, the applied load is measured using strain-gauge sensors KISTLER 108524-5500®, placed below the springs of the load cells (10, Fig. 3.1; 9, Fig. 3.2) and capable to operate up to 500 N with a sensitivity of 0.1 N.

An eddy current sensor MICRO-EPSILON eddyNCDT3010® is placed at the end of the metal cart platform to evaluate the sliding amplitude by measuring the variation of the magnetic field between a fixed plate and the extremity of the sliding carter (10, Fig. 3.2). Such sensor allows constant measurements with a sensitivity of 0.1 μm.

The oscillating frequency of the metal carter holding the specimens is given by the servo electric engine KEB TA41® (5, Fig. 3.1) and it can be regulated by a dedicated controller KEB COMBIVERT F5.

Friction force, which is the most complicated variable to be measured, is evaluated by a tangential force detecting device. A pressing mechanism, consisting of rigid horizontal rods (11, Fig. 3.2), encloses the vertical arms bearing the counter-specimens. In this way, the tangential component of the contact load, deriving from the friction existing between specimens and counter-specimens, is transmitted to strain-gauge sensors BURSTER 85041® (11, Fig. 3.1; 12, Fig. 3.2). Considering all the losses in the load transmission between the contact areas and the sensors, knowing the measured tangential and the applied loads, the friction force at the contact areas can be determined.

Temperature is measured by type K thermocouples placed in both chambers and in the connecting pipes (17, Fig. 3.1). The amount of oxygen in the molten Pb/Pb alloy is monitored by yttrium stabilised zirconium sensors (16, Fig. 3.1). Level sensors, which take advantage of the conductivity of liquid metals/alloys, are placed in different positions in the facility to give information about the level of the molten Pb/Pb alloy in the two chambers of the facility (18, Fig. 3.1).

The sensor signals are visualised and stored on a PC through a program created with Lab VIEW® (Laboratory Virtual Instrument Engineering Workbench). This program received the signal through a I/O controller by WAGO (750-341)® consisting in several digital/analog slots connected to the corresponding sensors.

3.4 Oxygen control and liquid metal handling

According to [178] and in agreement with Eq. 3.1 and Eq. 3.2, in the FRETHME chambers, the oxygen concentration (C_O) in the liquid Pb/Pb alloy is attained and kept constant by controlling the oxygen partial pressure (pO_2) in the above atmosphere through the H_2/H_2O ratio

Eq.3.1.

$$p_{O_2} = \frac{p_{H_2O}^2}{p_{H_2}^2} = \exp\frac{2\Delta_f G^\circ(H_2O)}{RT}$$

In Eq.3.1, p_i is the partial pressure of the vapour component i, $\Delta_f G^\circ$ is the Gibbs free energy of formation, T is the temperature in Kelvin and R is the Boltzmann constant. At the thermodynamic equilibrium between the gaseous and the liquid phase (HLM), the concentration of oxygen dissolved in liquid Pb/Pb alloy depends on the oxygen partial pressure of the above atmosphere.

In Eq.2 γ is the chemical activity coefficient of the oxygen, a_O and c_O are the oxygen chemical activity and concentration in the molten metal/alloy, p_{O2} is the oxygen partial pressure in the atmosphere above the molten phase and the suffix s refers to the saturation conditions.

Eq.3.2. $$a_0 = \gamma_0 c_0 = \frac{c_0}{c_{0,s}} = \left(\frac{p_{O_2}}{p_{O_2,s}} \right)^{1/2}$$

As explained in Chapter 1, in order to facilitate the growth of an adequate protective oxide scale, the oxygen content in liquid Pb must be kept in the range 10^{-6} - 10^{-8} wt%.

The gas supply designed and installed to control the oxygen content in liquid Pb alloy is depicted in Fig. 3.3. It consists of two 50 l gas bottles, one containing argon and one argon + 5% hydrogen with 200 and 150 bar respectively. After reduction of the pressure to 1.5 bar the gases pass separately through a filtering system and a mass flow controller that determines the effective gas injection per unit of time (ml/min) in the system. Through the mixing of the two gases it is possible to control the partial pressure of H_2 in the Ar+H_2 gas, which goes in a distilled water bottle inside a thermostat. The gas, after bubbling through the distilled water at defined temperature, flows through a glass wool layer aimed to avoid water drops reaching the outgoing gas pipe. Since the partial pressure of the water is determined by the thermostat temperature it is simple to establish exactly the H_2/H_2O ratio necessary to have the desired oxygen partial pressure in the atmosphere above the molten Pb/Pb alloy. The moisturized gas moisture is injected in the FRETHME facility through small pipes welded to the chambers (15, Fig. 3.1).

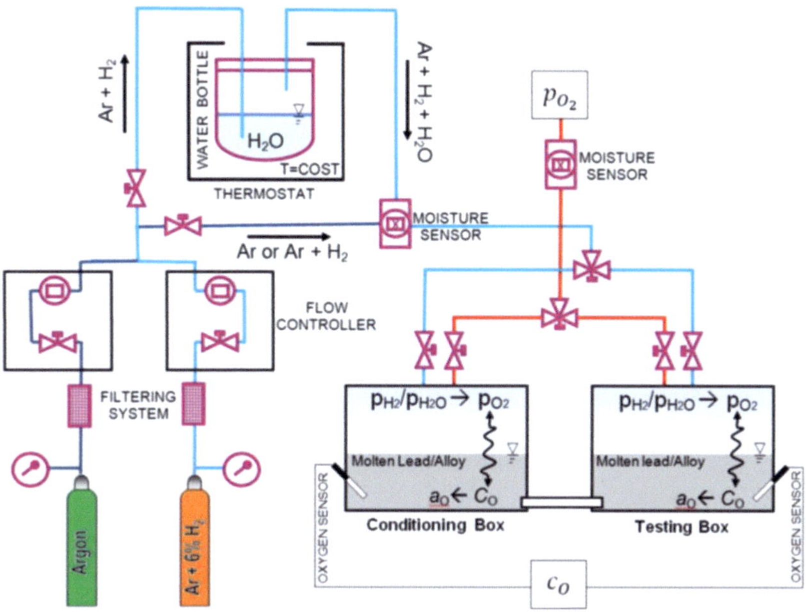

Fig.3.3. Sketch of the gas supply apparatus of the FRETHME facility. In the picture, *a* and *c* are the oxygen chemical activity and concentration in the molten metal/alloy, *p* is the partial pressure in the atmosphere above the molten phase.

For a rigorous control of the oxygen content in the chambers of the facility high tightness of the whole apparatus (facility chambers, connections, gas supply system etc.) is necessary. With respect of this, the chambers lid and all the movable parts crossing the chambers walls (friction force detecting apparatus, sliding carter and counter-specimen vertical arms) are sealed by GARLOCK Helicoflex® sealing for high temperature applications and appositely produced for the facility. Such sealing are rings of Ni-alloy Inconel 600 covered by an aluminium sheath and able to withstand up 1000 bar.

To transfer the molten Pb alloy from the conditioning chamber to the testing chamber and vice versa, pure Ar is pressed with 1-2 bar in the chamber. Likewise, when it is necessary to change the molten metal/alloy (due to oxidation, impurities and solution of alloying elements), the liquid can be pumped out from the facility chambers through clearing valves (13-14, Fig. 3.1).

Picture of the FRETHME facility are shown in Fig. 3.4 and 3.5.

Fig.3.4. Picture of the FRETHME facility from the top.

Fig.3.5. Picture of the FRETHME facility from the side.

4. Experimental activity

4.0 Introduction

The experimental activity at the base of this work consisted of three main actions. (i) The realization and installation of a facility for fretting test in HLMs, the FRETHME facility described in Chapter 3. (ii) The execution of a preliminary set of fretting tests in air, which were mainly aimed to confirm the wear mechanisms and validate the reliability and suitability of the FRETHME facility as device for fretting test. Besides, the fretting tests in air could be used to create a specific database for comparison with the experiments afterwards performed in liquid lead. (iii) The execution of fretting tests in liquid lead at high temperature and controlled amount of oxygen in solution.

Starting from some designers indications for components of lead cooled nuclear systems, fretting tests were performed in molten Pb varying the main fretting affecting variables (e.g. load, amplitude and number of cycles) and the temperature, which strongly influences the corrosion mechanisms in HLMs.

Moreover, different materials, chosen among the candidate materials for SG and in-core components of LFRs and ADSs, were tested under different fretting conditions.

Several characterization methods (profilometry and microscopy) were used to determine the fretting wear damage and the interaction between fretting and corrosion mechanisms occurring in liquid Pb.

The motivation of this work is to provide to designers experimental evidence about the conditions, in terms of fretting variables, which minimized the fretting processes (e.g. corrosion and wear) and their effect on the components performances and lifetime. As a consequence, in the experimental campaign, temperature, load and amplitude were varied to find the most adequate values for complying with lead cooled nuclear systems requirements (e.g. components lifetime and operating conditions).

4.1 Materials and Methods

4.1.1 Specimens and counter-specimens

The materials selected for the fretting tests are the ferritic-martensitic steel 9Cr1Mo T91, the austenitic Ti stabilized 15-15Ti steel (1.4970) and a surface modified T91 (GESA-T91). The compositions of these alloys, measured by EDX analysis, are reported in Tab. 4.1. The material for the specimens was provided as 1 m pipes afterwards resized to adequate dimensions by a Corundum/SiC rotating wheel blade. The nominal external diameter is 8.5 and 9.5 mm for T91 and 15-15Ti pipes respectively, while the wall thickness is 500 μm for both.

The counter-specimens were extracted from massive blocks of T91 and 15-15Ti by EDM (Electro Discharge Machining). The counter-specimens are 2 mm thick plates, shaped in order to have a maximum contact surface of 8 mm^2. Dimensional details of specimens and counter-specimens are shown in Fig. 4.1.

Material	Cr	Mo	Ni	Al	Ti	Y	Si	Mn	C	Nb	V
T91	8.26	0.95	0.13	-	-	-	0.43	0.38	0.105	0.075	0.2
GESA-T91	11-12	-	0.2	6-7	-	0.3	0.19	0.4	-	0.6	0.15
15-15Ti	15	0.48	15	-	0.48	-	0.3	1.40	0.11	-	-

Tab. 4.1
Measured composition of the alloys used for fretting tests in molten lead [179, 181]. In the Table Fe is balance.

The surface modified T91 (GESA-T91) specimens were obtained by a two step procedure carried out before the pipes resizing. First, at Sulzer Metco in Switzerland, FeCrAlY powders were deposed on a substrate of T91 pipe by LPPS (Low Pressure Plasma Spray). The deposed layer of FeCrAlY powders is characterized by high porosity, high roughness (up to average roughness Rq of 15 μm) and scarce adhesion to the T91 substrate (Fig. 4.2 – left). To improve these aspects, a second step, consisting in the GESA electron beam treatment, was performed at our laboratories. As a result of such treatment, the deposed layer melts with the first 5-10 μm of the T91 substrate. Thus, the surface modified T91 (GESA-T91) is characterized by a surface alloyed layer of few tens of micrometers, which contains aluminium and is characterized by a better surface finish (average roughness Rq around 5 μm) and adhesion to the substrate than after LPPS. Such features are highlighted in Fig. 4.2, where a SEM picture in BSE mode of a cross-section of a surface modified T91 specimen after LPPS and after GESA treatment is shown. Additionally, Fig. 4.3 reports the concentration of the alloying elements along the specimen cross-section before and after the electron beam treatment (GESA). The reduction of Al content after GESA treatment is caused by evaporation due to the high temperature locally reached during such a process, necessary for melting the deposed powders with the substrate.

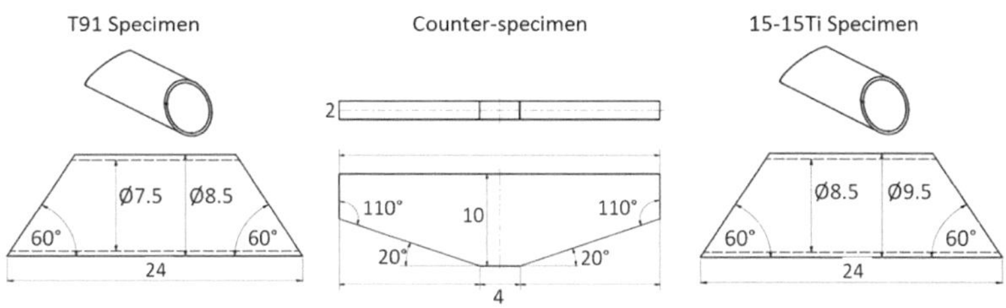

Fig.4.1. Dimensional details of specimens and counter-specimen. Lengths are given in mm.

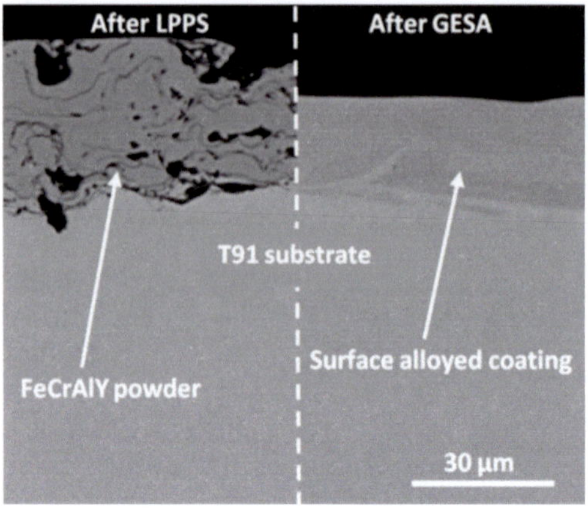

Fig.4.2. SEM picture in BSE mode of surface modified T91 cylinder after LPPS deposition of FeCrAlY and after electron beam treatment by GESA.

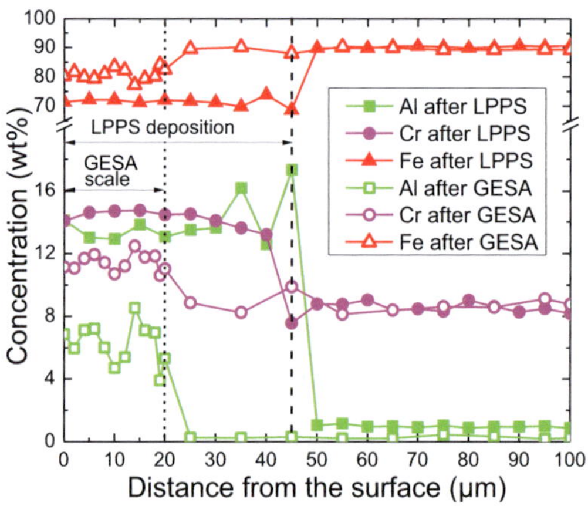

Fig.4.3. Concentration of the main alloying elements vs. distance from the surface, after LPPS and after GESA treatment.

4.1.2 Analysis methods and fretting damage assessment

After fretting tests, specimens and counter-specimens were extracted from the facility. In case of tests conducted in air, the fretting debris accumulated in the surrounding of the contact area were collected and analysed. After the fretting corrosion test carried out in molten lead, specimens and counter-specimens were cleaned from adhered Pb by chemical etching for 24 h with a solution 1:1:1 of: C_2H_5OH (Ethanol), CH_3COOH (Acetic acid) and H_2O_2 (Hydrogen peroxide).

The fretted surfaces were characterized by Light optical microscopy (LOM) using the microscopes Leitz Aristomet and Olympus BX60H. Specimens and counter-specimens were in addition investigated by the scanning electron microscopes (SEM) Hitachi S800 and Philips XL40.

The composition and the distribution of the alloying elements were evaluated by energy dispersive x-ray spectrometry (EDX) attached to the microscopes. The mentioned characterization methods were applied both on the fretted surface and on the fretting areas cross-sections, attained by cutting with a Corundum/SiC rotating wheel blade.

Microstructure analyses were also executed on specimens cross-sections by LOM and SEM. For this purpose, a specific etching procedure of about one minute was executed. A solution 3:3:1 of H_2O (distilled water), HCl (Hydrochloric acid) and HNO_3 (Nitric acid) was used for the 15-15Ti steel. A solution 8:1:1 of C_2H_5OH (Ethanol), HCl (Hydrochloric acid) and HNO_3 (Nitric acid) with the addition of 1 g of $C_6H_3N_3O_7$ (Picric acid) every 100 ml of solution was used for T91 steel and GESA-T91.

The morphology of the fretted surface and fretting wear damage in terms of material volume loss and wear depth were evaluated by a 3D light profilometry technique using the NanoFocus μScan Explorer. In particular, for every friction contact (couple specimen/counter-specimen), the fretting wear damage was calculated as volume loss (sum of the volume of material removed from the specimen and from the counter-specimen) and maximum fretting depth (measured on the specimens). From these values, other parameters descriptive of the fretting wear such as wear rate and specific wear coefficient were calculated (see details in chapter 5).

4.2 Experimental plan

A preliminary experimental campaign was carried out in air to evaluate the reliability of the FRETHME facility equipment in terms of: functioning of mechanical parts and sensors, heating and gas supply system, data acquisition and reproducibility of the results. Additionally, tests in air were used as reference to compare with fretting tests performed in molten lead at high temperature to evaluate possibly the role of molten lead in the fretting process. Fretting tests in air were also important to determine an adequate analysis route for the correct assessment of the fretting damage. Fretting experiments in air were performed mainly on T91 steel, for a total number of 22 friction pairs (specimen + counter-specimen), listed in Tab. 4.2.

The objective of the experimental campaign carried out in molten lead was to investigate the impact of the main affecting parameter on the fretting process and related mechanisms for the selected materials. For this purpose different applied loads, sliding amplitudes, temperatures, exposure time/number of cycles were tested. Fretting test details are given in Tab. 4.3 where the experimental plan is reported. The fretting tests were performed with friction pairs (specimen/counter-specimen) of T91/T91 (30), 15-15Ti/15-15Ti (19) and GESA T91/T91 (9) for a total number of 58 friction pairs.

Exp.	Material (spec/counter)	Temp. (°C)	$[O_2]$ conc. (wt%)	No. of Cycles	Load [N]	Ampl. [µm]	Freq. [Hz]
1	T91/T91	25	atm	$5 \cdot 10^5$ 10^6 $1.72 \cdot 10^6$ $5 \cdot 10^6$ 10^7	50	75	20
2	T91/T91	25	atm	$1.72 \cdot 10^6$	12.5 25 50 60 75 100	75	20
3	T91/T91	25	atm	$1.72 \cdot 10^6$	50	25 50 75 100	20
4	T91/T91	25	atm	$1.72 \cdot 10^6$	50	75	10 20 30 40
5	T91/T91 15-15Ti/15-15Ti	300	atm	$5.4 \cdot 10^6$	50	75	10
6	T91/T91 15-15Ti/15-15Ti	450	atm	$5.4 \cdot 10^6$	50	75	10

15-15Ti/15-15Ti

Tab.4.2.

Fretting test performed in air

Exp.	Material (spec./counter)	Temp. (°C)	[O] conc. (wt%)	No. of cycles	Time (h)	Load (N)	Ampl. (µm)	Freq. (Hz)
7	T91/T91 T91/T91 T91/T91	450	10^{-6}	$5.4 \cdot 10^{6}$	150	50	75	10
8	T91/T91 T91/T91 T91/T91	450	10^{-6}	$5.4 \cdot 10^{6}$	150	50	75	10
9	T91/T91 15-15Ti/15-15Ti GESA T91/T91	450	10^{-6}	$5.4 \cdot 10^{6}$	150	50	75	10
10	T91/T91 15-15Ti/15-15Ti GESA T91/T91	450	10^{-6}	$1.12 \cdot 10^{6}$	312	50	75	10
11	T91/T91 15-15Ti/15-15Ti GESA T91/T91	450	10^{-6}	$2.16 \cdot 10^{7}$	600	50	75	10
12	T91/T91 15-15Ti/15-15Ti GESA T91/T91	450	10^{-6}	$3.35 \cdot 10^{7}$	930	50	75	10
13	T91/T91 T91/T91 T91/T91	450	10^{-6}	$5.4 \cdot 10^{6}$	150	25 50 75	75	10
14	15-15Ti/15-15Ti 15-15Ti/15-15Ti 15-15Ti/15-15Ti	450	10^{-6}	$5.4 \cdot 10^{6}$	150	25 50 75	75	10
15	T91/T91 15-15Ti/15-15Ti	450	10^{-6}	$5.4 \cdot 10^{6}$	150	15	75	10
16	T91/T91 15-15Ti/15-15Ti GESA T91/T91	450	10^{-6}	$5.4 \cdot 10^{6}$	150	50	35	10
17	T91/T91 15-15Ti/15-15Ti GESA T91/T91	450	10^{-6}	$5.4 \cdot 10^{6}$	150	50	165	10

18	T91/T91 15-15Ti/15-15Ti GESA T91/T91	450	10^{-6}	$5.4\cdot10^{6}$	150	50	15	10
19	T91/T91 15-15Ti/15-15Ti	450	10^{-6}	$5.4\cdot10^{6}$	150	15	190	10
20	15-15Ti/15-15Ti 15-15Ti/15-15Ti 15-15Ti/15-15Ti	450	10^{-6}	$5.4\cdot10^{6}$ $1.12\cdot10^{7}$ $3.35\cdot10^{7}$	150 312 930	50	230	10
21	T91/T91 T91/T91 T91/T91	450	10^{-6}	$5.4\cdot10^{6}$	150	15 25 75	35	10
22	T91/T91 T91/T91 T91/T91	450	10^{-6}	$5.4\cdot10^{6}$	150	15 25 75	165	10
23	T91/T91 15-15Ti/15-15Ti GESA T91/T91	500	10^{-6}	$5.4\cdot10^{6}$	150	50	75	10
24	T91/T91 15-15Ti/15-15Ti GESA T91/T91	550	10^{-6}	$5.4\cdot10^{6}$	150	50	75	10
25	T91/T91 15-15Ti/15-15Ti	450	10^{-6}	$5.4\cdot10^{6}$	150	50	75	10
26	T91/T91 T91/T91 T91/T91	450	10^{-6}	$5.4\cdot10^{6}$	150	50	75	10

Tab.4.3

Fretting tests performed in molten lead

4.3 Considerations on the fretting contact

To evaluate correctly the fretting damage and to compare fretting test results among the specimens of different selected materials, especially in terms of wear damage, an analysis of the contact geometry was done. Indeed, some important factors like surface of contact, contact pressure and total removable material, depend on the contact geometry. It was evaluated whether the slightly smaller diameter of the T91 pipes compared to the 15-15Ti pipes has any influence on the fretting contact geometry.

In Fig. 4.4 a 3D CAD representation of the contact specimen/counter-specimen is shown and in Fig. 4.5 the scheme of the specimen/counter-specimen contact cross-section occurring during fretting tests in the FRETHME facility is illustrated.

The breakthrough of the cladding/pipe is assumed as limit for the fretting damage. As a consequence, the maximum depth of the fretting damage will be 500 μm, being this the pipes wall thickness.

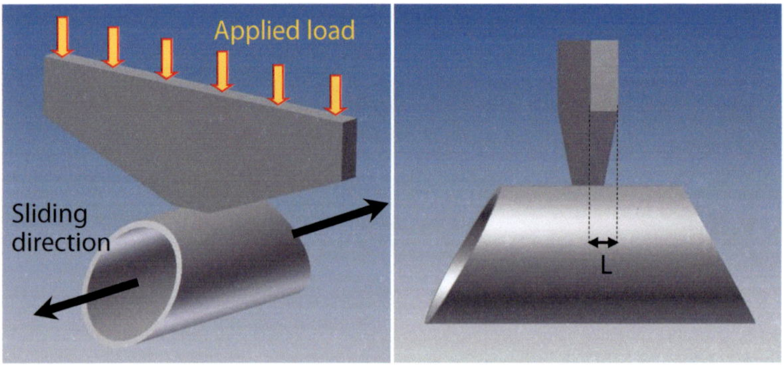

Fig.4.4. 3D CAD representation of the contact specimen/counter-specimen in the FRETHME facility.

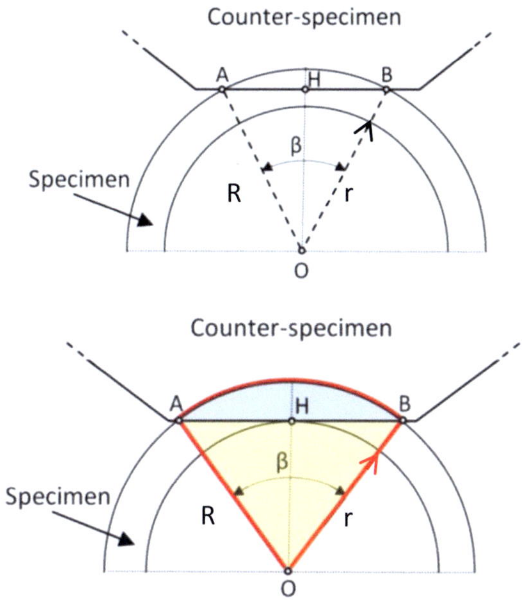

Fig.4.5. Cross-section scheme of the contact between specimen and counter-specimen in the FRETHME facility.

In Fig. 4.5, a certain depth of the fretting damage identified by the position *H* corresponds to a certain segment *AB*, which is the transversal length of the fretting contact that determine the contact surface. Being *R* and *O* the outer radius and the centre of the pipe respectively and *β* the central angle of the circular sector determined *O*, *A* and *B*, with simple trigonometric calculations contact surface (*S*), contact pressure (*P*) for a given applied load and total removable volume (V) can be calculated.

Eq.4.1 $$S = \overline{AB} \cdot L$$

L is the thickness of the counter-specimen (2 mm), which determines the longitudinal length of the fretting contact. $\overline{AB}$ is twice the distance $\overline{AH}$, and it can be calculated as follows:

Eq.4.2 $$\overline{AB} = 2 \cdot \overline{AH} = 2 \cdot \left(R \cdot sin\left(\frac{\beta}{2}\right)\right)$$

With β/2 deriving from:

Eq.4.3 $$\cos\left(\frac{\beta}{2}\right) = \frac{\overline{OH}}{R} \rightarrow \frac{\beta}{2} = \cos^{-1}\left(\frac{\overline{OH}}{R}\right)$$

And with: $R - \overline{OH} = fretting\ depth \rightarrow \overline{OH} = R - fretting\ depth$

So that the contact surface can be obtained as shown in Eq. 4.4:

Eq.4.4 $$S = 2 \cdot L \cdot \left(R \cdot sin\left(cos^{-1}\left(\frac{\overline{OH}}{R}\right)\right)\right)$$

The maximum/minimum value of the segment $\overline{OH}$ corresponds to the outer/inner radius of the cylinder. Thus, neglecting possible plastic deformation, and knowing radius and thickness of T91 and 15-15Ti pipes, it is possible to calculate contact surface and the contact pressure deriving from a certain applied load. In Fig. 4.6 and Fig. 4.7 the evolution of contact surface and pressure with the wear depth ($R - \overline{OH}$) during the fretting process are plotted for the materials under investigation. For the surface modified T91, an outer radius of 4.27 mm is considered due to the additional 20 µm of surface alloyed coating resulting from LPPS and GESA treatment.

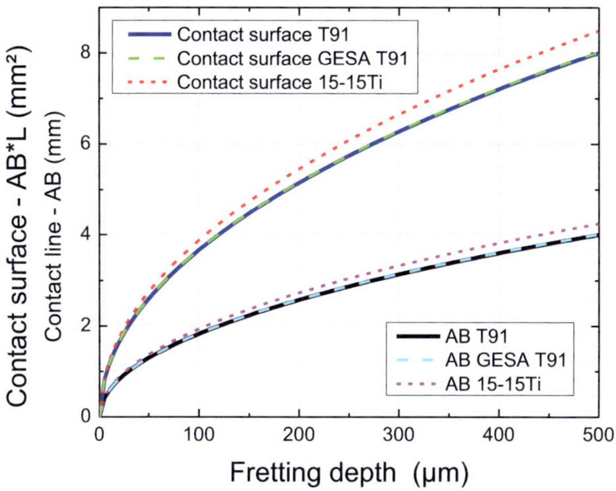

Fig.4.6. Evolution of the contact area and of the line AB (Fig. 4.5) with the depth of the fretting movement.

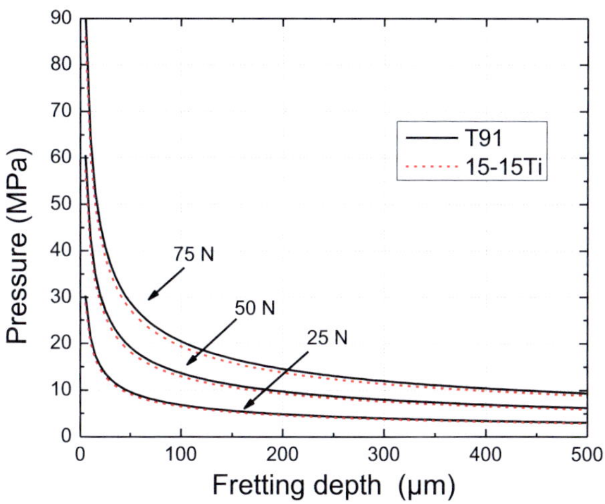

Fig.4.7. Evolution of contact pressure for a fixed applied load with the depth of the fretting movement.

A fuel cladding pipe penetration of 10% can be arbitrarily taken as the safety margin for the maximum allowed fretting depth. Such limit, considering the dimensional details shown in Fig. 4.1, corresponds to a maximum fretting depth of 50 µm. Fig. 4.6 and 4.7 highlight that for fretting depths of this magnitude, the influence of the geometrical differences between the specimens is absolutely negligible. Indeed, at a fretting depth of 50 µm, for T91, 15-15Ti and GESA-T91, contact line and contact surface are 1.3 / 1.375 / 1.303 mm and 2.6 / 2.749 / 2.606 mm² respectively.
If the limit condition is considered the breakthrough of the pipe ($\overline{OH}$ = r), as reported in Eq. 4.5, the maximum volume *V* of material removable during the fretting process (Eq. 4.5) is given by the circular segment defined by the chord $\overline{AB}$ (light blue area in Fig. 4.5) multiplied by the counter-specimen thickness (*L*). The circular segment can be calculated as the difference between the area of the circular sector AÔB (red line Fig. 4.5) and the area of the triangle $\widehat{AOB}$ (yellow area Fig. 4.5).

Eq.4.5
$$V = \left(\frac{\pi \cdot R^2 \cdot \beta}{360} - \frac{\overline{AB} \cdot \overline{OH}}{2}\right) \cdot L$$

Applying Eq. 4.2 and Eq. 4.3 to Eq. 4.5 and the known values of inner and outer radius for the different specimens, the total removable volume can be calculated according to Eq. 4.6.

Eq.4.6
$$V = \left(\frac{\pi \cdot R^2 \cdot \beta}{360} - \cdot \left(R \cdot sin\left(cos^{-1}\left(\frac{r}{R}\right)\right)\right) \cdot r\right) \cdot L$$

Using Eq. 4.6 the maximum removable volume for the three materials under study can be calculated. It is 2.69, 2.7 and 2.83 mm³ for T91, GESA-T91 and 15-15Ti, respectively (< 10% difference). Such small difference is further reduced at the maximum allowed fretting depth of 50 µm. In this case the volume removable from T91, GESAT91 and 15-15Ti is 0.084, 0.88 and 0.085 mm³ respectively (< 5 % difference).

This simple analysis suggests that it is reasonable to assume that the narrow differences of radius of the three different selected materials will not have any remarkable influences on the fretting wear. Indeed, such difference does not lead to significant variation in contact pressure ($\leq 5\%$) or total material removable during the fretting process. Thus, it can be assumed that fretting wear damages affecting specimens of different materials are directly comparable.

4.4 Results reliability and errors

Every scientific measurement, as far as precise and accurate, is subjected to uncertainties, which, if not properly considered, might favour misleading judgements or conclusions. Thus, a rigorous scientific activity cannot be regardless of estimating and analysing the measurement errors.
In this work, two main eventual sources of errors were indentified: random errors in test results due to the fretting test execution and systematic errors in fretting wear damage quantification by profilometry. Such sources of uncertainties are analysed separately in the following paragraphs.

4.4.1 Statistic analysis of fretting results

Random errors or deviations in fretting test results, expressed as volume loss or depth of the fretting damage, might be caused by different factors. Indeed, repeating several fretting tests at the same conditions and parameters, rarely exactly identical results are obtained.
Some uncontrollable variations, such as uncorrected (e.g. not completely normal) alignment of the contact pair specimen-counter-specimen and/or small changing of some fretting parameters (e.g. load, amplitude, frequency, temperature and oxygen content), might occur leading to more or less significant variation in the test result.

Assuming that the same kinds of variations can possibly happen regardless from materials and conditions tested, the statistic analysis of result variations for repeated fretting tests provides an instrument to verify the reliability of the results and to determine the uncertainty to be considered.

For this purpose, 10 experiments with the same setup were performed. T91 was chosen as materials for specimens and counter-specimens and the testing conditions, listed in Tab. 4.4, were selected according to the most recurring setup in the experimental campaign.

Temperature (°C)	[O] conc. (wt%)	number of cycles	Time (h)	Load (N)	Amplitude (µm)	Frequency (Hz)
450	10^{-6}	$5.4 \cdot 10^{6}$	150	50	75	10

Tab.4.4

Fretting t parameters of the experiments executed for the statistic analysis.

As shown in Tab. 4.5 there is a certain scatter of the values of volume loss measured after repeated fretting tests. These values are distributed around a mean value ($\bar{x} = 0.0304$) with a certain standard deviation ($\sigma_x = 0.00227$) that corresponds to a standard deviation < 10 % ($\sigma_x = 7.5\%$). Thus, FRETHME facility does allow to obtain a good reproducibility of the results.

<table>
<tr><th>Test (N)</th><th>Result (x_i) [mm^3]</th><th>Mean value ($\bar{x}$)</th></tr>
<tr><td>1</td><td>0.032</td><td rowspan="3">$\bar{x} = \sum_{i=1}^{N} x_i \Big/ N = 0.0304$</td></tr>
<tr><td>2</td><td>0.035</td></tr>
<tr><td>3</td><td>0.030</td></tr>
<tr><td>4</td><td>0.027</td><td rowspan="2">Standard deviation (σ_x)</td></tr>
<tr><td>5</td><td>0.032</td></tr>
<tr><td>6</td><td>0.029</td><td rowspan="5">$\sigma_x = \sqrt{\frac{1}{N-1} \cdot \sum_{i=1}^{N} (x_i - \bar{x})^2} = 0.00227$</td></tr>
<tr><td>7</td><td>0.030</td></tr>
<tr><td>8</td><td>0.031</td></tr>
<tr><td>9</td><td>0.030</td></tr>
<tr><td>10</td><td>0.028</td></tr>
<tr><td></td><td></td><td>Percentage error ($\sigma_x\%$) = 7.5%</td></tr>
</table>

Tab.4.5

Data for the statistic analysis of the fretting test results

4.4.2 Underestimation in white light profilometry

To evaluate the fretting wear damage, a 3D profilometer is used. A scheme of the working principle of such technique is shown in Fig. 4.8. A white light beam is focused on the surface to analyse through a lens with chromatic length aberration. Due to this aberration, the focus point of a specific wavelength corresponds to each Z-position. The light hitting the surface is reflected and sent to a spectrometer through a pin hole. The spectrometer provides an intensity curve depending on the wavelength and the focused wavelengths is the one corresponding to the maximum of intensity.

The procedure adopted to evaluate the fretting damage consists in surface profilometry scanning of the fretted surface and surrounding area.

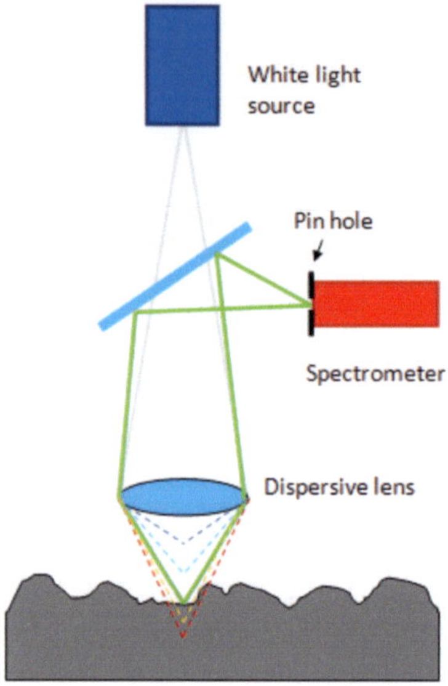

4.8. Scheme of the functioning of the 3D profilometer.

As a result, the topography of the fretted specimen is attained (Fig. 4.9) and, through a dedicated software, information about the fretting wear, such as volume of removed material, fretted area extension, depth of the fretting damage and roughness parameters, are achieved.

As detailed in Chapter 2, during the fretting process, the formation of an interposing layer made of compacted debris is not unusual in the contact area. Such scale, despite of the fact that it is still adhered to the specimen surface (in the fretting grooves), should be considered as part of the fretting damage. Indeed, as it will be shown in the results chapter, the compacted scale is characterized by different chemical, mechanical and physical properties compared to the underlying material and it can be generally considered inadequate to withstand the reactor operating requirements. Thus, to avoid underestimation of the fretting wear damage, the compacted scale formation must be taken into account when the data resulting from the profilometry analysis are collected. Indeed, as shown in Fig. 4.10 for a T91 specimen, the profilometry scan does not recognize the compacted scale, which generally varies between 10 and 20 µm. Thus, if the volume of material removed is intended to be used as indicative of the fretting damage, a safety margin that accounts for the thickness of the compacted scale should be added to the measured volume loss. Alternatively, the maximum fretting depth can be adopted as parameter for fretting damage estimation because it was noticed that it generally corresponds to an area without compacted scale.

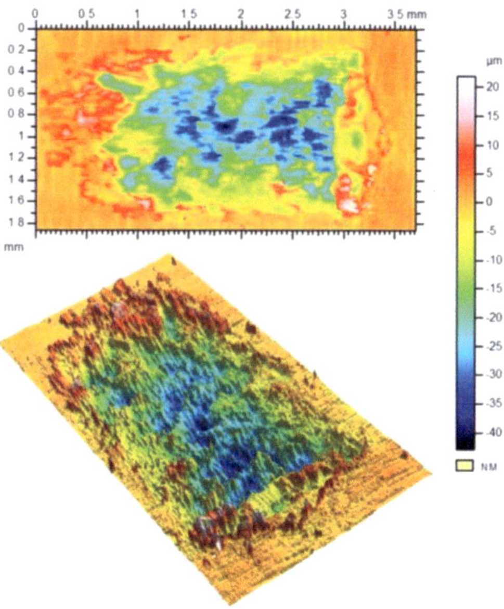

Fig.4.9. Example of imagine obtained through white light profilometry scanning analysis.

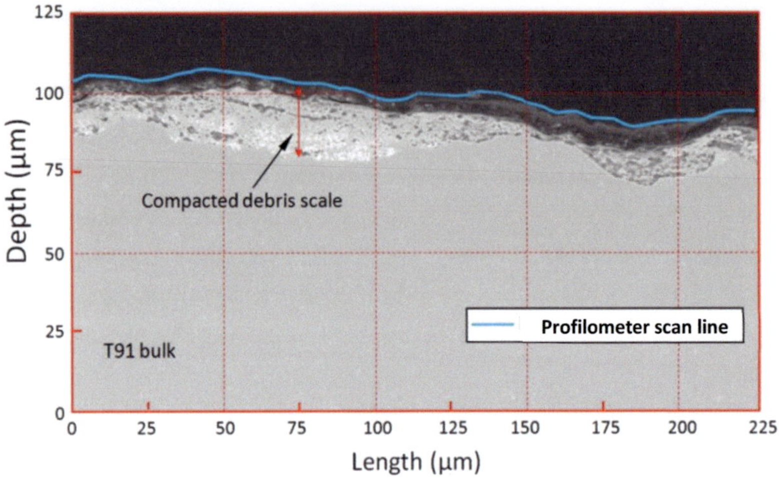

Fig.4.10. Overlapping of profilometry scanning line and SEM picture in BSE mode of the same cross-section, which belongs to a T91 specimen after fretting test in molten lead for 150 h, $5.4 \cdot 10^6$ cycles, 75 N, 165 µm and 10 Hz.

4.5 Results representation

The results of fretting tests are discussed in the following chapters focusing mainly on two particular aspects: the fretting wear damage and the fretting corrosion effects.

The fretting wear damage is evaluated by means of laser profilometry and expressed using four different parameters: volume loss (V), fretting depth (D), wear rate (W) and specific wear coefficient (K).

The volume loss (V) is the sum of the volume removed by fretting action from specimen and counter-specimen.

The fretting depth (D) corresponds to the maximum depth of the fretting groove that is measured on the tested specimens. Additionally, to evaluate the seriousness of the fretting damage with respect to a real application, the fretting depth is also expressed as penetration percentage of a real fuel cladding tube.

The wear rate (W) is the volume loss per time unit (mm^3/s). Experimentally accessible wear rates are the average rates which are calculated by the overall volume loss per overall experimental time.
The specific wear coefficient (K) is the volume loss normalized with respect to total sliding distance and applied load. It is calculated according to the Archard's equation of wear as follow [182]:

Eq. 4.7 $$V = K \cdot N \cdot D \rightarrow K = \frac{\mathrm{V}}{\mathrm{N \cdot D}} = \frac{V}{N \cdot n \cdot 2 \cdot A}$$

In Eq. 4.7, V is the volume lost during the fretting process in m^3, N is the applied load in [N], and D is the total sliding distance in [m] that is given by twice the product of sliding amplitude (A) and number of cycles (n). The factor 2 is added to the divisor because each fretting cycle is twice the sliding amplitude. As a result K is expressed in $m^3 \cdot N^{-1} \cdot m^{-1}$ or alternately in Pa^{-1}.
Due to the normalization, the specific wear coefficient can be used to compare tests carried out in different experimental conditions.
In addition, the evolution fretting wear process is represented through 3D images and cross-section profiles resulting from the profilometry analysis.
Fretting corrosion effects are presented and discussed, especially for fretting tests in liquid lead.

5. Fretting tests in air: results and discussion

5.1 Results

5.1.1 General appearance of the fretting damage

The nature of the fretted area, except some difference of composition, is similar for the friction pairs tested. The peculiarities noticed during the analysis are common for friction pairs of both T91 and 15-15T that were the materials tested in air.

The friction between specimen and counter-specimen during the fretting process results in a sort of ploughing action that causes the formation of a groove on the fretted surfaces. The material removed forms fretting debris, which can be metallic or oxidized, and that are scattered by the fretting motion in the surrounding of the fretted area in form of reddish powder. In the SEM micrograph of Fig. 5.1 typical fretting debris is shown. The debris particle size varies between fractions of µm and 2-3 µm.

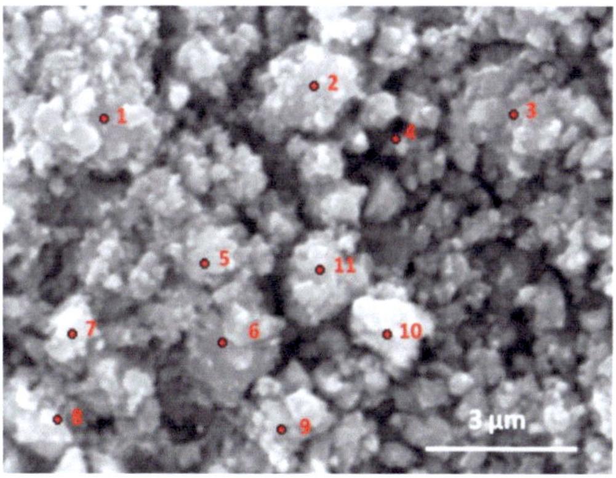

Fig.5.1. SEM micrograph of fretting debris formed during fretting test at room temperature in air. Test parameters: $1.72 \cdot 10^6$ cycles, 75 µm, 50 N, 20 Hz.

According to an EDX analysis performed at marked points of Fig. 5.1 the fretting debris consist mainly of Fe-oxides (Tab. 5.1).

(wt%)	O	Cr	Fe
1	37.3	4.5	56.1
2	34.6	4.7	59.5
3	30.7	4.8	61.8
4	29.2	5.1	63.2
5	31.1	5.0	62.3
6	35.1	4.8	57.4
7	39.4	4.2	53.7
8	26.6	5.5	65.4
9	28.5	4.9	63.4
10	36.6	4.4	57.4
11	28.9	4.8	63.5

Tab.5.1. Data of EDX analysis performed in points marked by red numbers in Fig. 5.1.

The fretted area generally presents a reduction in the specimen cross-section due to the material removal (Fig. 5.2). Besides, as shown in the specimen cross-section profile of Fig. 5.2, at the border of the fretted area material accumulation might occur.

Moreover, on the surface of the fretted area, the formation of an interposing scale made of compacted oxidized debris is commonly observable. Thickness and extension of the compacted scale vary with the test parameters. However, formation of interposing compacted scales thicker than 10 μm is unusual. The SEM/EDX line scan analysis reported in Fig. 5.3 shows and confirms the presence of an interposing scale of about 8 μm made of compacted oxidized debris.

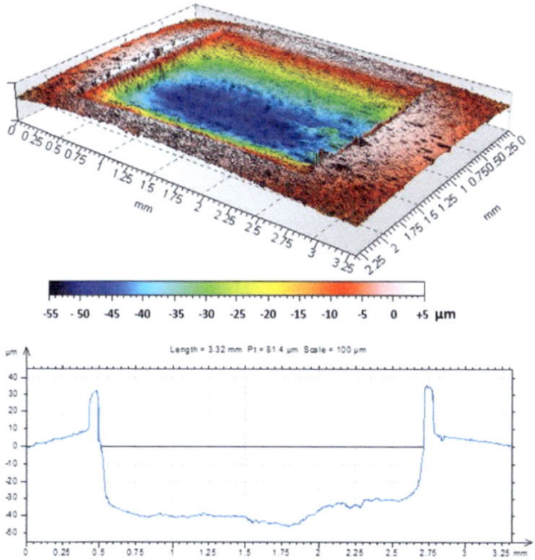

Fig.5.2. 3D profilometry and cross-section profile of a T91 specimen after $1.728 \cdot 10^6$ fretting cycles in air at 25°C. Test parameters: 100 µm, 50N, 20 Hz.

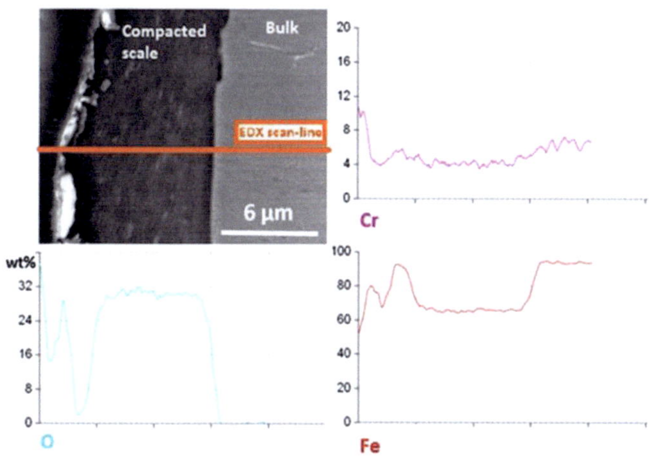

Fig.5.3. SEM/EDX line scan analysis (orange line) of a T91 specimen cross-section after test in air at 25°C. Test parameters: $1.72 \cdot 10^6$ cycles, 75 µm, 75 N, 20 Hz.

5.1.2 Influence of exposure time/number of cycles

To evaluate the evolution of the fretting process with the time/number of cycles, tests were performed with fixed values of applied load (50 N), sliding amplitude (75 μm) and sliding frequency (20 Hz). The number of fretting cycles was varied between $5 \cdot 10^5$ and 10^7 (corresponding to test duration of about 7 to 140 h).

Fretting wear, in terms of volume loss and fretting depth, increases with the number of cycles (full and empty squares in Fig. 5.4). In particular, as shown in Fig. 5.4, a faster increase of the volume loss during the first $\sim 2 \cdot 10^6$ cycles (run-in period) is followed by a slower increase of the curve with a nearly linear dependency on the number of cycles. After 10^7 fretting cycles, volume removed from the friction pair and maximum depth measured are 0.188 mm^3 and 89 μm (17.8 % of penetration) respectively. Fig. 5.4 also shows that the trend of volume loss and fretting depth are very similar.

The observations made so far are confirmed by the wear rate plot reported in Fig. 5.5 (empty circles). The run-in period is characterized by a wear rate of about $1.5 \cdot 10^{-6}$ mm^3s^{-1} that remains roughly constant up to 10^6 cycles. For higher number of fretting cycles, the mean wear rate decreases down to a minimum of $3 \cdot 10^{-7}$ mm^3s^{-1} after 10^7 cycles. The plot of the specific wear coefficient (Fig. 5.5), after a quasi constant trend around $1 \cdot 10^{-14}$ $m^3N^{-1}m^{-1}$ during the run-in period, decreases with the number of cycles down to a minimum of about $2 \cdot 10^{-15}$ $m^3N^{-1}m^{-1}$after 10^7 cycles.

As highlighted by the fretted area cross-sections profiles shown Fig. 5.6, increasing the number of cycles the fretting depth globally increases and the formation of two depressions at the sliding turning points becomes more and more evident.

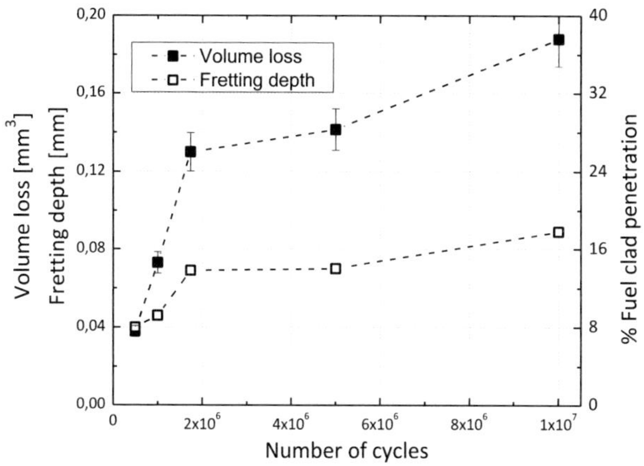

Fig.5.4. Graph of volume loss and fretting depth versus number of cycles for T91 tested in air at room temperature. Test parameters: 50 N, 75 μm, 20 Hz.

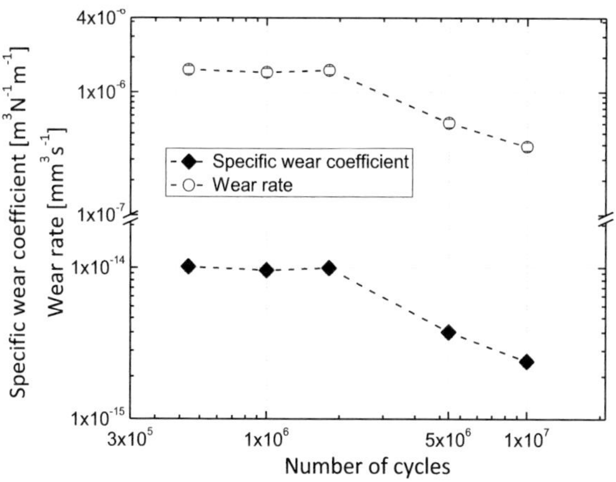

Fig.5.5. Graph of wear rate and specific wear coefficient versus number of cycles for T91 tested in air at room temperature. Test parameters: 50 N, 75 μm, 20 Hz.

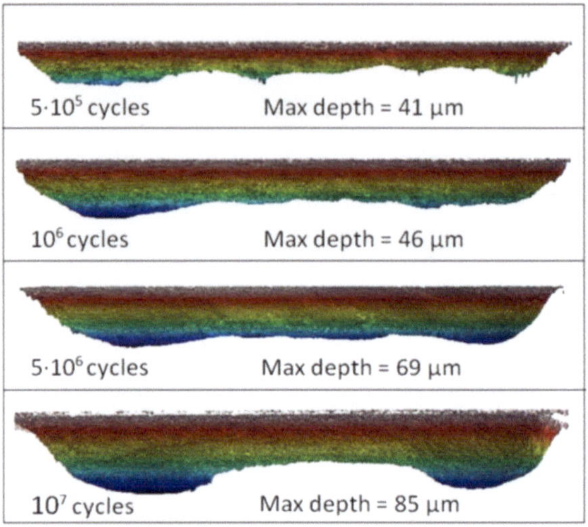

Fig.5.6. (left) Laser profilometry analysis of the fretting grooves of T91 specimen submitted to fretting tests of different duration in air at room temperature. Test parameters: 50 N, 75 μm, 20 Hz.

5.1.3 Influence of applied load

The influence of the applied load on the fretting process was investigated performing tests with a fixed number of cycles ($1.72 \cdot 10^6$), sliding amplitude (75 μm) and sliding frequency (20 Hz). Different loads were applied in the range from 12.5 to100 N.

In Fig. 5.7, volume loss and fretting depth increase with the increasing applied load up to a turning point, around 50 N, and then decrease.

The minimum value of volume loss and fretting depth are 0.018 mm³ and 5 μm respectively for an applied load of 100 N. The maximum volume loss and fretting depth are 0.13 mm³ and 69 μm (13.8 % of penetration) respectively, resulting from the test performed with an applied load of 50 N.

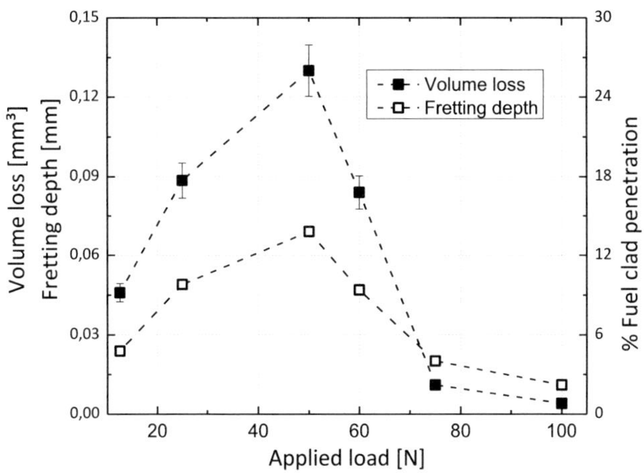

Fig.5.7. Graph of volume loss (full squares) and fretting depth (empty squares) versus applied load for T91 tested in air at room temperature. Test parameters: $1.72 \cdot 10^6$ cycles, 75 µm, 20 Hz.

In Fig. 5.8, wear rate and specific wear coefficient are depicted. The wear rate (empty circles in Fig. 5.8) increases with the load up to 50 N where the maximum value of about $1.5 \cdot 10^{-6}$ mm^3s^{-1} was measured. For higher loads the wear rate gradually decreases down to a minimum of about $4.5 \cdot 10^{-8}$ mm^3s^{-1} corresponding to an applied load of 100 N.

Up to 50 N, the specific wear coefficient (full diamonds in Fig. 5.8) is almost constant around the value of $\sim 1 \cdot 10^{-14}$ $m^3 \cdot N^{-1} \cdot m^{-1}$; whereas for higher loads it decreases down to a minimum of $\sim 1 \cdot 10^{-16}$ $m^3 \cdot N^{-1} \cdot m^{-1}$ corresponding to an applied load of 100 N.

The maximum and minimum values of wear rate and specific wear coefficient differ by almost two orders of magnitude.

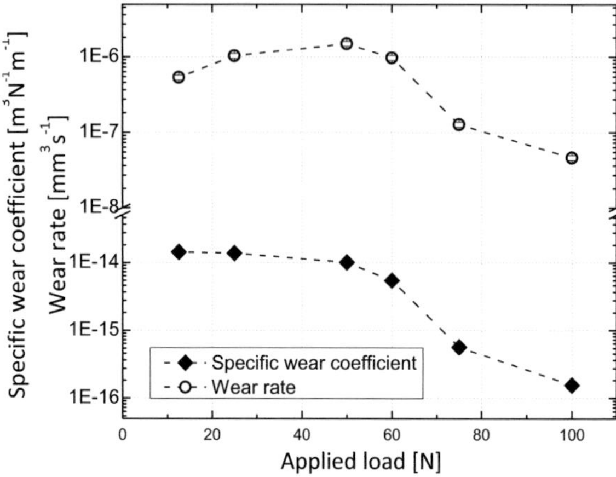

Fig.5.8. Graph of wear rate (empty circles) and specific wear coefficient (full diamonds) versus applied load for T91 tested in air at room temperature. Test parameters: $1.72 \cdot 10^6$ cycles, 75 µm, 20 Hz.

Variations in the applied load influence the morphology of the fretted surface. The SEM micrograph in Fig. 5.9 is illustrative for such changes in the morphology of the fretted area for friction pairs of T91 tested with different applied loads.

The image compares the appearance of the fretted areas resulting from fretting performed with applied load of 50 and 100 N (left and right side of Fig. 5.9 respectively). The fretting area (marked in Fig. 5.9 by the yellow dotted line) resulting from the test performed with a load of 50 N is characterized by debris accumulation at the groove borders (the turning points of the fretting motion). Besides, it is more extended than the one resulting from the test with 100 N, which on the other hand is characterized by compacted debris.

The SEM/EDX analysis and data shown in Fig. 5.10 and Tab. 5.2 respectively, confirmed the debris accumulation at the border of the fretted area after fretting test with applied load of 50 N. The EDX data, performed at points 1-10 in Fig. 5.10, support the presence of metal and oxidized debris (which is most probably Fe_2O_3 as indicated by the Fe/O ratio).

The SEM image of Fig. 5.11 shows the fretted area cross-section of a specimen tested with applied load of 100 N. It is possible to notice a scale of compacted debris (dark grey area in Fig. 5.11) up to about 10 μm thick that forms on the steel fretted area. The outcomes of the EDX analysis performed in the points identified by red numbers in Fig. 5.11 and reported in Tab. 5.3, suggests that the compacted scale is mainly made of oxidized debris. The Fe/O ratio suggests the presence of Fe_2O_3.

Specimens tested with applied load smaller than 60 N, showed similar characteristics (e.g. debris accumulation at the border) to the one tested with 50 N. On the other hand the specimen tested with applied load of 75 N showed similar characteristics like the one tested with 100 N.

Fig.5.9. SEM micrographs of the fretted areas resulting from fretting test performed on T91 in air, at room temperature and with applied load of 50 N (left) and 100 N (right). Test parameters: $1.72 \cdot 10^6$ cycles, 75 μm, 20 Hz.

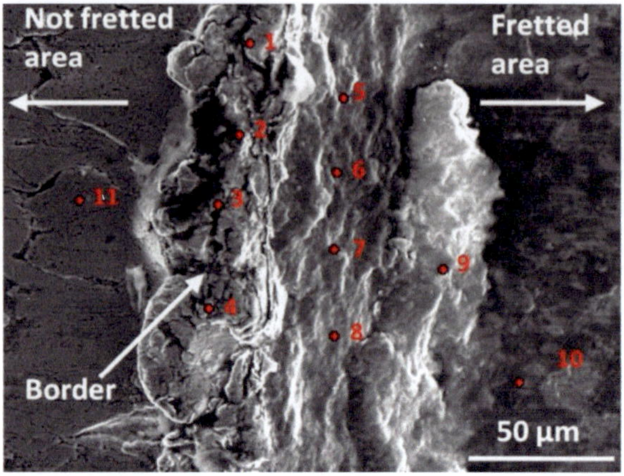

Fig.5.10. SEM micrograph of the debris accumulation at the border of the fretted area resulting on T91 after test in air at room temperature and applied load of 50 N. Test parameters: $1.72 \cdot 10^6$ cycles, 75 µm, 20 Hz.

(wt%)	O	Cr	Fe
1	2.1	4.4	91.8
2	2.8	6.0	90.0
3	6.9	7.6	84.7
4	27.1	0.6	72.2
5	27.6	6.0	65.4
6	24.0	6.2	68.8
7	30.7	5.8	62.3
8	19.4	6.8	72.6
9	30.3	5.9	62.4
10	15.6	6.8	76.8
11	0.5	4.6	93.9

Tab.5.2. EDX data of points marked by red numbers in Fig. 5.10.

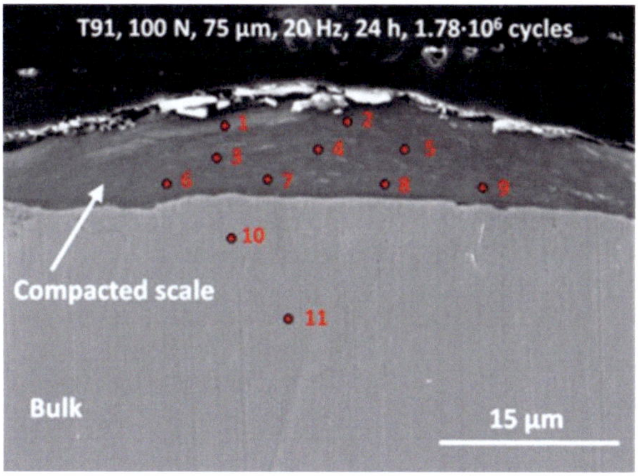

Fig.5.11. SEM image of the cross-section of the fretted area resulting from fretting of T91 in air at room temperature with applied load of 100 N. Test parameters: $1.72 \cdot 10^6$ cycles, 75 µm, 20 Hz.

(wt%)	O	Cr	Fe
1	19.8	6.7	73.4
2	7.0	7.1	85.8
3	28.6	5.5	65.7
4	29.3	5.5	65.1
5	23.3	5.7	70.8
6	25.2	5.7	69.0
7	26.3	5.7	67.9
8	26.3	6.2	67.4
9	25.5	6.0	68.3
10	0.4	8.5	91.6
11	0.6	8.9	89.9

Tab.5.3. EDX point analysis performed in the points marked by red numbers in Fig. 5.11.

5.1.4 Influence of the amplitude

The role of the sliding amplitude in the fretting process was studied performing tests with a fixed number of cycles ($1.72 \cdot 10^6$), applied load (50 N) and sliding frequency (20 Hz). The different sliding amplitude was varied in the range from 25 to 100 μm.

Volume loss and fretting depth increase with increasing sliding amplitude as shown by Fig. 5.12.. The minimum value of volume loss and fretting depth are 0.0023 mm³ and 13 μm respectively, measured after the fretting test with sliding amplitude of 25 μm. The maximum values recorded were 0.213 mm³ and 94 μm (18.8% of penetration), resulting from the test performed with amplitude of 100 μm. Minimum and maximum values differ by almost 2 orders of magnitude for volume loss and by a factor 7 for fretting depth. A remarkable change in the curves trend can be noticed for amplitudes larger than 50 μm.

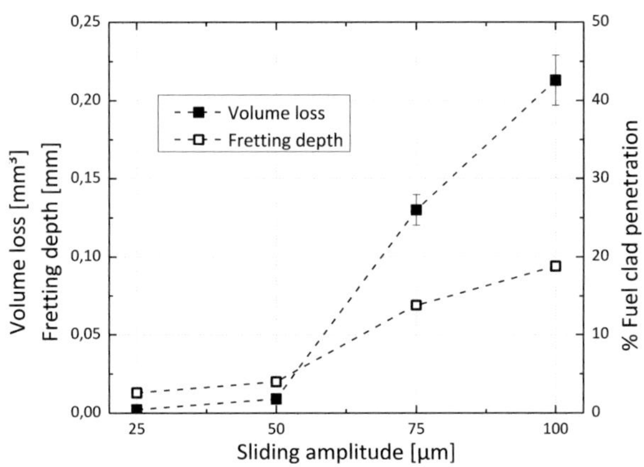

Fig.5.12. Graph of volume loss and fretting depth versus sliding amplitude for T91 tested in air at room temperature. Test parameters: $1.72 \cdot 10^6$ cycles, 50 N, 20 Hz.

The observations made so far find confirmation in Fig. 5.13, where wear rate and specific wear coefficient are plotted for different sliding amplitudes. Between 25 to 75 µm, the wear rate (Fig. 5.13) increases from ~$2\cdot10^{-8}$ to ~$1\cdot10^{-6}$ mm^3s^{-1} (around 2 orders of magnitude); whereas for sliding amplitude of 100 µm, the wear rate only increases up to ~$2\cdot10^{-6}$ mm^3s^{-1}.

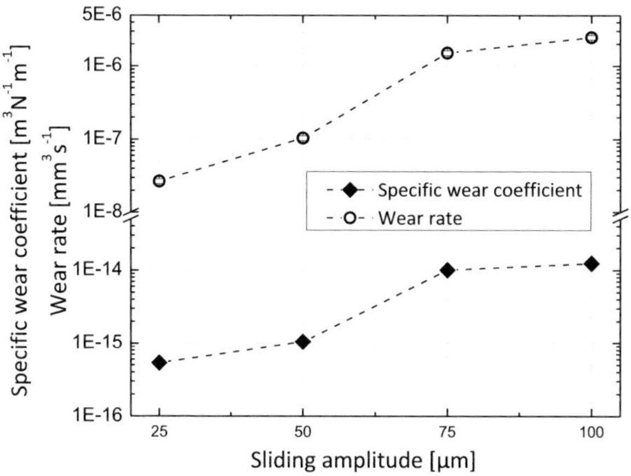

Fig.5.13. Graph of wear rate and specific wear coefficient versus sliding amplitude for T91 tested in air at room temperature. Fretting parameters: $1.72\cdot10^{6}$ cycles, 50 N, 20 Hz.

The specific wear coefficient (Fig. 5.13) follows the same trend. It increases from ~$5\cdot10^{-16}$ to ~$1\cdot10^{-14}$ $m^3N^{-1}m^{-1}$ from 25 and 75 µm; whereas it remains nearly constant for amplitudes larger than 75 µm.

In Fig. 5.14, the fretting wear area (delimited by the yellow dotted line) resulting from the test performed with sliding amplitude of 100 µm (left side of Fig. 5.14) is broader than the one resulting from the fretting test with amplitude of 50 µm (right side Fig. 5.14). Moreover, with amplitude of 100 µm, fretting debris accumulation at the border of the fretted area can be noticed.

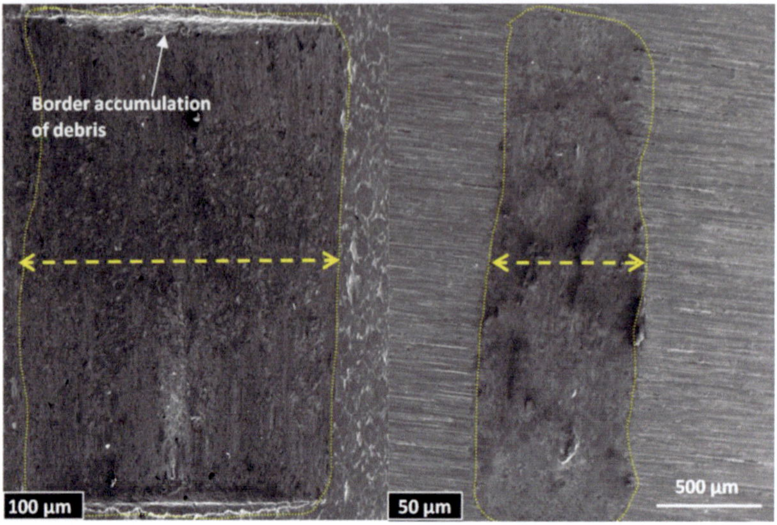

Fig.5.14. SEM micrograph of the fretted areas resulting from fretting test performed on T91 in ait at room temperature with sliding amplitude of 100 µm (left) and 50 µm (right). Test parameters: $1.72 \cdot 10^6$ cycles, 50 N, 20 Hz.

Fig. 5.15 shows SEM imaging at high magnification of the fretted area resulting from tests with amplitude of 100 µm (left side) and 50 µm (right side). As visible in this picture and as confirmed by the EDX point scan data reported in Tab. 5.4, with 100 µm amplitude a significant portion (> 50 %) of the fretted area in not covered by compacted debris. EDX analysis performed in points 1-4 highlight the normal T91 composition, while points 5 and 6 reveal the presence of debris. On the contrary, with 50 µm amplitude, compacted oxidized debris cover most of the fretted area (points 7-11 Tab. 5.4).

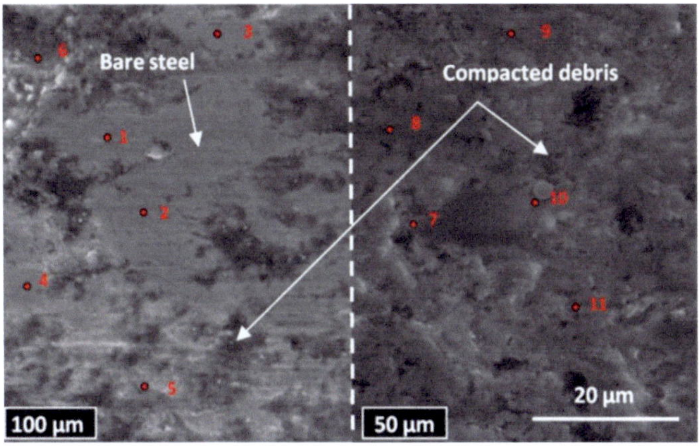

Fig.5.15. SEM micrograph at high magnification of the fretted area observed on T91 after test performed in air at room temperature with sliding amplitude of 100 µm and 50 µm. Test parameters: $1.72 \cdot 10^6$ cycles, 75 µm, 20 Hz.

(wt%)	O	Cr	Fe	
1	1.2	8.4	89.2	100 µm amplitude
2	1.0	8.5	89.7	
3	0.8	8.1	90.5	
4	0.0	8.2	90.5	
5	23.6	6.7	68.9	
6	26.3	6.6	66.4	
7	37.3	5.8	56.1	50 µm amplitude
8	31.4	6.1	62.0	
9	36.1	5.5	57.7	
10	36.6	5.3	57.4	
11	0.5	4.6	93.9	

Tab.5.4. EDX point analysis performed in the points marked by red numbers in Fig. 5.15.

5.1.5 Influence of the frequency

The effect of the sliding frequency on the fretting process was analysed performing tests with a fixed number of cycles ($1.72 \cdot 10^6$), applied load (50 N) and sliding amplitude (50 μm). Different sliding frequencies were imposed between 10 and 40 Hz.

As visible in Fig. 5.16, up to 30 Hz the volume loss (full squares) and the fretting depth (empty squares) are only slightly influenced by variations in sliding frequency. From 10 to 30 Hz the volume loss and the fretting depth change from 0.0085 to 0.0125 mm^3 and from 13 to 23 μm respectively. However, a further increase of the frequency up to 40 Hz leads to a significant increase (by a factor 5) up to 0.062 mm^3 of the volume loss.

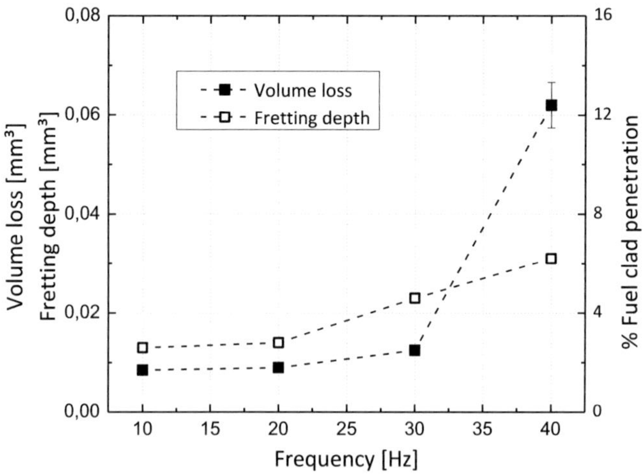

Fig.5.16. Graph of volume loss (full squares) and fretting depth (empty squares) versus frequency for T91 tested in air at room temperature. Test parameters: $1 \cdot 10^6$ cycles, 50 μm, 50 N.

The wear rate and specific wear coefficient curves given in Fig. 5.17 agree with these observations. The wear rate (empty circles) is much higher at 40

Hz (1 order of magnitude or more) than for frequency lower that 30 Hz. Similar, between 10 and 30 Hz the specific wear coefficient (full diamonds) is nearly constant and around an order of magnitude lower than at 40 Hz (~$1 \cdot 10^{-14}$ $m^3N^{-1}m^{-1}$).

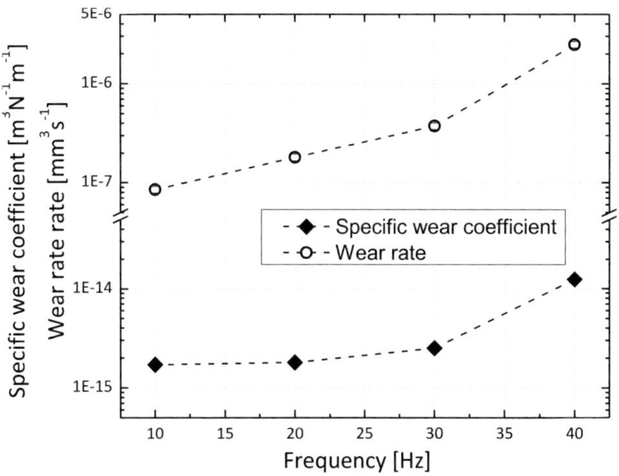

Fig.5.17. Graph of wear rate (empty circles) and specific wear coefficient (full diamonds) versus frequency for T91 tested in air at room temperature. Test parameters: $1 \cdot 10^6$ cycles, 50 μm, 50 N.

With sliding frequency of 40 Hz the scale of compacted debris has a different composition compared to the one detected for lower frequencies (e.g. in Fig. 5.3). Indeed, as shown in the SEM/EDX point analysis of Fig. 5.18 and Tab. 5.5, together with nearly pure Fe-oxide (points 1-6) traces of Fe-Cr mix oxides, most probably spinel type (points 7-9), are detected in the scale of compacted debris.

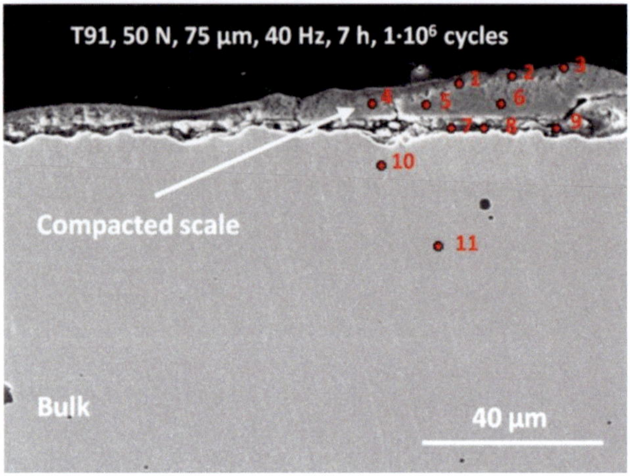

Fig.5.18. SEM micrograph of the compacted scale formed on T91 in air at room temperature with a frequency of 40 Hz. Test parameters: $1{\cdot}10^6$ cycles, 50 µm, 50 N, 40Hz.

(wt%)	O	Cr	Fe
1	22.5	0.9	76.5
2	26.9	0.8	72.2
3	29.0	0.6	70.3
4	27.1	0.6	72.2
5	32.6	0.9	66.4
6	29.6	1.0	69.3
7	18.7	40.0	41.2
8	23.7	18.3	57.9
9	31.9	23.6	44.3
10	0.3	8.5	91.7

Tab.5.5. EDX point analysis performed in the points marked by red numbers in Fig. 5.18.

5.1.6 Influence of the temperature

The effect of the temperature was investigated comparing results collected from fretting tests at room temperature with results of tests performed on friction pairs of T91 and 15-15Ti at 300 and 450°C with comparable setup (e.g. sliding amplitude, number of cycles and applied load).

Volume loss and fretting depth measured on friction pairs of T91 and 15-15Ti after fretting test in air at room temperature, at 300°C and at 450°C, are plotted in Fig. 5.19. For a fixed number of fretting cycles, the volume loss decreases with increasing temperature both for T91 (squares) and 15-15Ti (triangles). For friction pairs of T91, from room temperature (25°C) to 300°C, the volume loss decreases by ~31% (from 0.141 to 0.098 mm^3) and from 300 to 450°C a further decrease by ~34% (down to 0.065 mm^3) is measured. For friction pairs of 15-15Ti the volume loss at 300°C is lower than the one of T91 and it decreases from 0.061 to 0.036 mm^3 (-31%) from 300 to 450°C.

The fretting depth (empty squares and triangles in Fig. 5.19) follows a similar trend. For T91 specimens, from room temperature to 300°C, the fretting depth decreases by ~36% (from 70 to 45 μm) and from 300 to 450°C, a further decrease by ~12% (down to 40 μm) is noticed. For 15-15Ti, from 300 to 450°C fretting depth passes from 31 to 26 μm (-~17%).

Both for T91 and 15-15Ti, wear rate and specific wear coefficient (Fig. 5.20) decrease remarkably when increasing the temperature from 300 to 450°C. For T91 from room temperature to 300°C, wear rate and specific wear coefficient remain nearly constant.

As shown in Fig. 5.21 for 15-15Ti, a compacted scale up to 30 μm thick forms in the fretted area. In Tab. 5.6, the EDX data indicates concentrations of Fe, Cr and O that suggest the presence of Fe oxides and Fe-Cr spinel oxides. For T91 similar results were noticed.

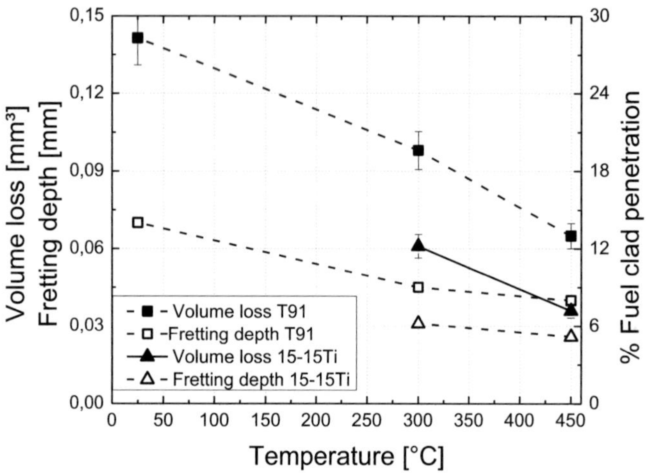

Fig.5.19. Graph of volume loss and fretting depth versus temperature for friction pairs of T91 (full and empty squares) and 15-15Ti full and (empty triangles). Test parameters: $5.4{\cdot}10^6$ cycles, 75 µm, 50 N, 10 Hz.

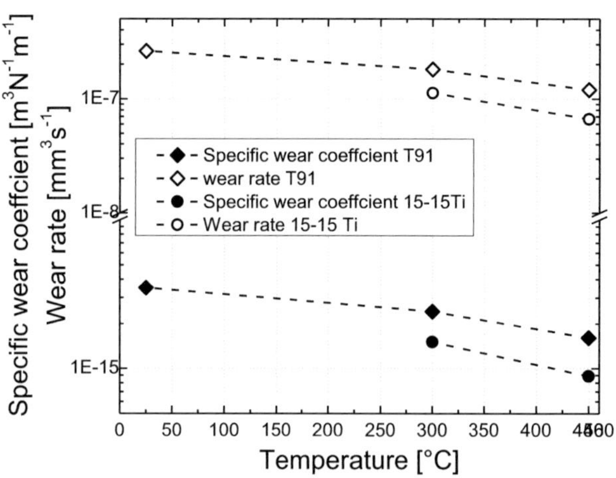

Fig.5.20. Graph of wear rate and specific wear coefficient versus temperature for friction pairs of T91 (full and empty diamonds) and 15-15Ti (full and empty circles). Test parameters: $5.4{\cdot}10^6$ cycles, 75 µm, 50 N, 10 Hz.

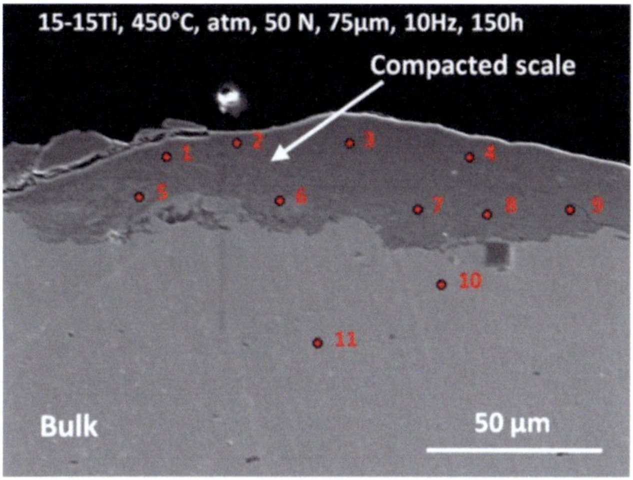

Fig.5.21. SEM micrograph of the compacted scale formed on a 15-15Ti specimen tested at 450°C.Test parameters: $5.4 \cdot 10^6$ cycles, 75 µm, 50 N, 10Hz, 150 h

(wt%)	O	Cr	Fe	Ni
1	29.20	10.47	49.71	10.62
2	27.99	11.49	48.11	12.41
3	26.50	12.32	49.22	11.95
4	26.85	12.10	48.89	11.16
5	28.85	12.44	50.82	8.89
6	28.52	11.37	48.92	11.22
7	27.49	11.66	49.07	11.89
8	28.93	11.57	50.16	9.34
9	28.94	10.87	50.53	10.66
10	0.30	15.35	69.12	15.23
11	0.16	15.37	65.32	15.15

Tab.5.6. EDX point analysis performed in the points marked by red numbers in Fig. 5.21

5.2 Discussion of the results

The results of the experimental campaign carried out in normal atmosphere at room temperature are generally in agreement with previous research activities and with the literature survey detailed in Chapter 2.

The analysis of friction pairs submitted to fretting test highlighted some of the fundamental characteristics of the fretting damage like the material removal and the formation the compacted interposing scale. Such a scale consists mostly of compacted oxides (oxidized debris) that, according to the Fe and O concentrations indicated by EDX analyses reported in paragraph 5.1.1, are most probably Fe_2O_3. This observation is compatible with the iron oxidation process in normal atmosphere at temperature below 200°C [130-137, 140].

Volume loss and fretting depth increase with the number of cycles according to a parabolic rate (Fig. 5.22). This outcome is in agreement with the literature [56, 121-123]. Indeed, as discussed in paragraph 2.3.1 for fretting in air, a quasi-linear time dependency, which follows an initial run-in period characterized by faster material removal, is common [66].

The initial higher wear rate is likely related to the removal of the oxide scale naturally covering steels. After a more severe period, the system evolves toward a steady state with an almost linear dependency on the number of fretting cycles (black dashed line in Fig. 5.22). Accordingly, for T91 the experimental outcomes showed a wear rate 2 to 4 times higher during the first 10^6 cycles (run-in period) than afterwards (Fig. 5.5). However, after 10^7 fretting cycles (168 h with a frequency of 20 Hz) the unlubricated fretting in air of T91 penetrates up to ~ 18 % of a real fuel clad thickness (Fig. 5.22).

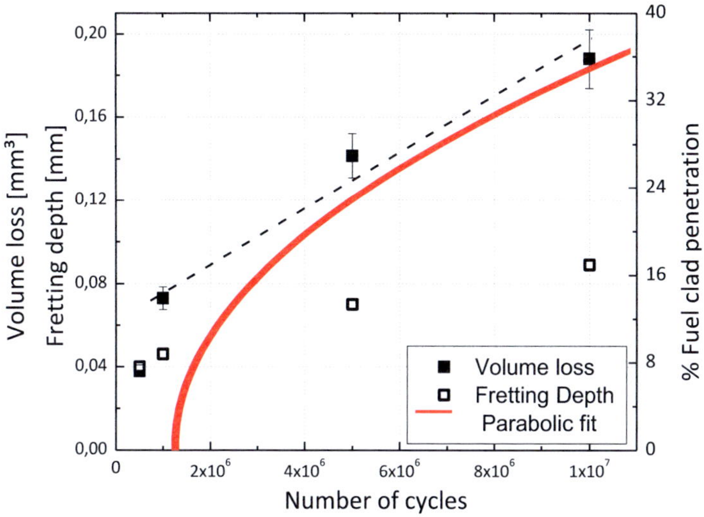

Fig.5.22 Graph of volume loss (full squares) and fretting depth (empty squares) versus number of cycles for friction pairs of T91 tested in air at room temperature. Test parameters: 50 N, 75 μm, 20 Hz.

As noticed in previous works [124] and described in paragraph 2.3.2, fretting wear in terms of volume loss and fretting depth increases with the applied load up to a maximum (turning point) and then decreases. Fretting tests performed in air with increasing applied load showed that for an imposed sliding amplitude of 75 μm, the maximum fretting wear is reached at 50 N (Fig. 5.7). For higher loads the volume loss gradually decreases down to a minimum for an applied load of 100 N. Wear rate and specific wear coefficient measured for the test performed with applied load of 50 N are almost 2 orders of magnitude higher than for the test carried out with 100 N (Fig. 5.8).

According to paragraph 2.4, the applied load together with the sliding amplitude play a central role in determining the fretting regime that defines specific contact conditions (slip, stick or mix stick-slip condition) and the predominant fretting process (wear, fatigue and corrosion) [63]. For constant sliding amplitude, increasing the applied load, the fretting regime can change together with the wear mechanism and the material removal rate (Fig. 2.16).

The specific wear coefficient (K), defined by Eq. 4.7 as the volume loss normalized with respect to the applied load and the total distance covered during the fretting process, can be used to compare the results of this work with the ones of previous research [63]. To obtain a fretting map as defined in paragraph 2.4, the specific wear coefficients calculated in this work are plotted in Fig. 5.23 as function of the applied load together with the analogous fretting map proposed by Vingsbo and Söderber in [63].

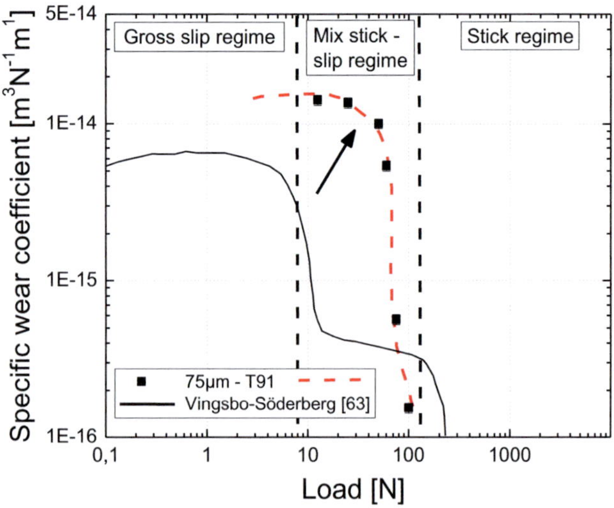

Fig.5.23. Fretting map of specific wear coefficient versus applied load for friction pairs of T91 tested in air at room temperature. Test parameters: 75 µm, 20 Hz, $1.73 \cdot 10^6$ cycles.

The comparison between the two data distributions highlights that the experimental data, although shifted to higher values (red dashed line), follow the trend reported in [63]. The difference between the two curves might be related to differences in material and amplitude (in [63] austenitic steels and amplitude of 90 μm are considered).

With an imposed sliding amplitude of 75 μm a change of fretting regime, from gross-slip to mix stick-slip, most probably occurs for an applied load around 50 N. For higher loads, the change of fretting regime causes a decrease of volume loss and fretting depth. Moreover, the significantly lower value of specific wear coefficient corresponding to an applied load of 100 N might suggest the beginning or the approaching of the stick regime.

The change of fretting regime, from gross slip to mix stick-slip regime, is also supported by SEM/EDX analyses (Figs. 5.9-5.11) that show thick and constant compacted scales covering the fretted area for high applied loads (75-100 N) and a partially uncovered steel surface for lower loads (15-50 N). It is reasonable to assume that the higher the load (and consequently the local pressure) the higher the debris confinement (retention) in the contact area. This probably favours the formation of thick (10-15 μm) compacted interposing scales that slow down the direct wear of steel and mitigate the fretting process (paragraph 2.2.4). On the contrary, low loads favour debris release and the formation of the interposing scale is slowed down (delayed).

Fretting tests performed in this work highlighted that fretting wear generally increases with the sliding amplitude. Clearly, increasing the amplitude also fretting path, fretting affected surface and as consequence also volume loss during the wear process increase. However, the increase of fretting wear with the amplitude is not merely related to an increase of the area affected by the wear process. Indeed, the specific wear coefficient, which contains an amplitude normalization of the fretting wear (Eq. 4.7), also increase with the amplitude (Fig. 5.13). If the amplitude dependency of fretting was only dimensional (higher amplitude = bigger damage area) specific wear coefficient and fretting depth (Fig. 5.12) would be the same for all the fretting tests performed at different amplitude. However, fretting depth and specific wear coefficient are significantly higher for tests performed with amplitude of 75/100 μm than for tests with amplitude of 25/50 μm. Therefore, variations in the imposed sliding amplitude seem to induce changing in the wear process, which probably means that there is a change of fretting regime.

Specific wear coefficients calculated according to Eq. 4.7 and reported in Fig. 5.24 as function of sliding amplitude confirm this observation.

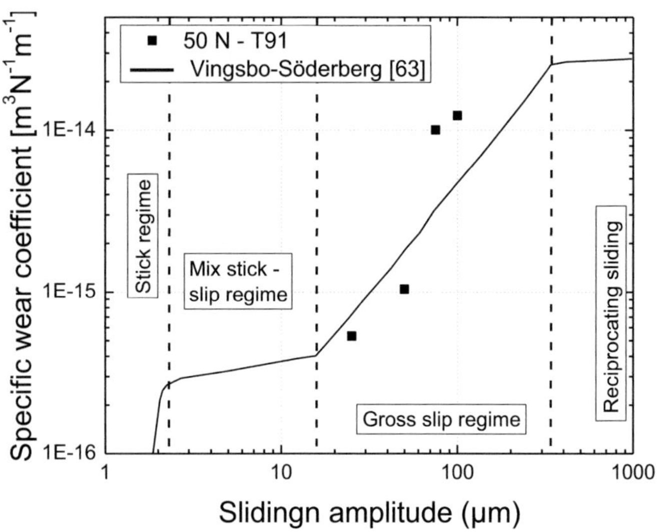

Fig.5.24. Fretting map of specific wear coefficient versus applied load for friction pairs of T91 tested in air at room temperature. Test parameters: 50 N, 20 Hz, $1.73 \cdot 10^6$ cycles.

In Fig. 5.24 it is possible to observe that the experimental data points follow the trends proposed by Vingsbo and Söderberg in [63] and plotted in this figure for comparison.

In the specific case of this work, for a fixed applied load of 50 N the change in fretting regime seems to take place for amplitudes around 50 μm. Indeed, for a larger amplitude sudden increase of fretting wear, wear rate and specific wear coefficient was noticed (Fig. 5.12 and 5.13). Besides, debris retention is more accentuated for amplitudes shorter than 50 μm than for larger amplitudes (e.g. 100 μm) (Fig. 5.14 and 5.15). This effect can be explained considering that shorter amplitudes favour debris retention in the contact area and formation of the interposing scale of compacted debris that mitigates the direct wear of the steel surface [71, 72, 89, 92, 106-114].

Differences between the fretting regimes border lines resulting from this work and from [63] are expected due to the different steels and applied loads investigated ([63] considers austenitic steels and loads in the range 1.5-34 N).

From the discussion on sliding amplitude and applied load it can be perceived that these two variables interplay closely in the fretting process. The different combinations of load and amplitude tested in this work determined the fretting regime and induced more or less severe fretting wear. It emerged that the combinations load / amplitude of 60-75-100 N / 75 μm and 50 N /25-50 μm likely favour the mixed stick-slip regime. The combinations load / amplitude of 12.5-25-50 N / 75 μm and 50 N / 75-100 μm favour the gross slip regime. Starting from these data (Fig. 5.23 and 5.24), with reference to paragraph 2.4 and Fig. 2.13, a fretting map load-amplitude can be created (Fig. 5.25). In this type of map, the plotted lines are the borders of the fretting regime domains that correspond to different load-amplitude combinations. In Fig. 5.25, each squared point corresponds to a combination load/amplitude. The colour of the square determines the order of magnitude of the specific wear coefficient and the fretting regime. The border lines (dashed lines) between gross slip and mix stick-slip regime and between mix stick-slip and stick regime were plotted considering 10^{-15} and 10^{-16} as order of magnitudes of the specific wear coefficient corresponding to mix stick-slip regime [63]. In Fig. 5.25 the dashed lines represents the shift of the fretting regime border lines compared to [63] (solid line).

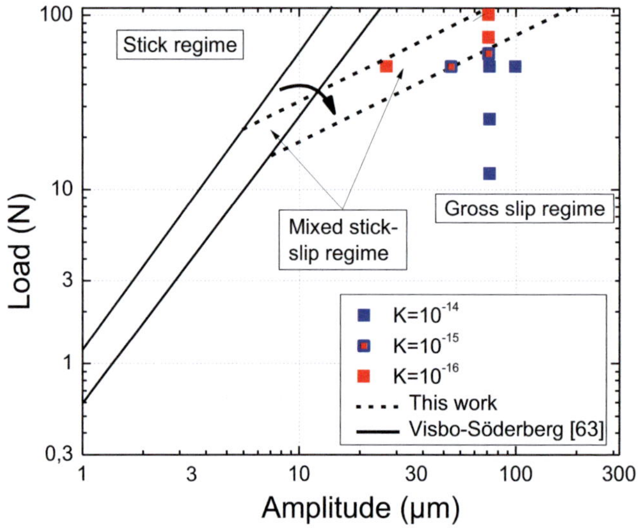

Fig.5.25 Fretting map sliding amplitude / applied load for friction pairs of T91 testes in air at room temperature. Test parameters: 25-100 µm, 12.5-100 N, 20 Hz, $1.73 \cdot 10^6$ cycles.

The role of sliding frequency was also investigated. Fretting wear barely increases with the frequency between 10 and 30 Hz, but at 40 Hz an increase of volume loss and specific wear coefficient by a factor 6 and by an order of magnitude was recorded.

The literature survey given in paragraph 2.3 [98, 102, 104, 128] pointed out that the role of frequency is still not totally understood. Some studies indicated that at lower frequencies fretting wear decreases; whereas for other studies the converse is true. However, two main aspects might be discussed to explain the results of this work. On the one hand, according to the delamination theory of wear (paragraph 2.2), fretting wear proceeds as nucleation-propagation of micro-cracks that cause the formation of the fretting debris. Such process is the result of a compression-traction stress field that is locally created by the cyclic loading (through the surface asperities) occurring during fretting. From this point of view, a higher frequency should lead to faster debris formation and consequently bigger wear. On the other hand, during the fretting process, the friction between specimen and counter-specimen causes a local temperature rise. The higher the frequency the higher the temperature rise in the contact area. The temperature increase favours the development of thicker oxide scales than at room temperature. An eventual detachment of the oxide scale would cause bigger volume removal and loss of the lubricating action provided by the oxide scale.

For friction pairs of T91 and 15-15Ti, the investigation of temperature effects on fretting in air indicated that fretting wear decreases with the increasing temperature (up to 450°C).

For T91, fretting tests carried out at 300 and 450°C highlighted e.g. a volume loss reduction compared to tests performed at room temperature with the same fretting parameters (e.g. load, amplitude and frequency), of 30 and 50 % respectively.

These experiments also indicated that friction pairs of T91 are affected more seriously by fretting wear (i.e. + ~ 35 % volume loss) than the ones of 15-15Ti (Fig. 5.19).

As discussed in [130-137, 140], up to 450-500 °C, fretting wear decreases with the temperature due to the formation of a stable oxide scale which acts as a solid lubricant and mitigates the fretting process. Around 200°C the oxide growth passes from following a logarithmic law to a parabolic one. This most probably justifies the decrease in fretting wear at 300°C. The further decrease of fretting wear at 450°C can be explained considering the different wear resistance properties of the predominant Fe-oxides forming at 300 and 450°C. Indeed, in normal atmosphere for temperature higher than 380°C the formation of magnetite (Fe_3O_4) is expected. Such Fe-oxide has higher wear resistance than hematite (Fe_2O_3) formed at lower temperatures [130-132].

The reason of the higher susceptibility to fretting wear in air of T91 steel compared to 15-15Ti might be related once again to the oxidation process. As explained in Chapter 1, due to composition (lower Cr content) and microstructural characteristics (smaller grain size) the martensitic T91 steel oxidized faster and more extensively than the austenitic 15-15Ti steel. As a result, T91 is generally covered by thicker oxide scales than 15-15Ti. Such oxide scale might be cracked and break off due to the fretting action. As consequence, T91 might be affected by greater volume loss than 15-15Ti.

To organize and compare different fretting conditions (tested) in one graph, the concept of working rate ($\dot{W}$) can be used [165]. Such parameter is related to the wear rate ($\dot{V}$) through the specific wear coefficient (K).

Eq. 5.1 $$\dot{V} = K \cdot \dot{W}$$

With $\dot{V} = \frac{V}{t}$ and $K = \frac{V}{N \cdot 2A \cdot n}$ (Eq. 4.7) → $\dot{W} = \frac{N \cdot 2A \cdot n}{t}$

In Eq. 5.1, $\dot{V}$ is the wear rate in m^3s^{-1}; $\dot{W}$ is a term containing load (*N*), amplitude (*A*) and number of cycles (*n*) expressed in m, total distance run by fretting (*2A·n*) and the experiment time (t) in s. $\dot{W}$ can be considered a work rate expressed in $N \cdot m \cdot s^{-1}$ or in watt (W). Since $\dot{W}$ includes number of cycles and time, it also intrinsically contains the frequency, which is determined by the number of cycles in the time (n/t [s] = frequency [Hz]).

The wear rate ($\dot{V}$) versus the work rate ($\dot{W}$) for all the experimental results is plotted in Fig. 5.26. Such graph shows the fretting wear severity for different conditions.

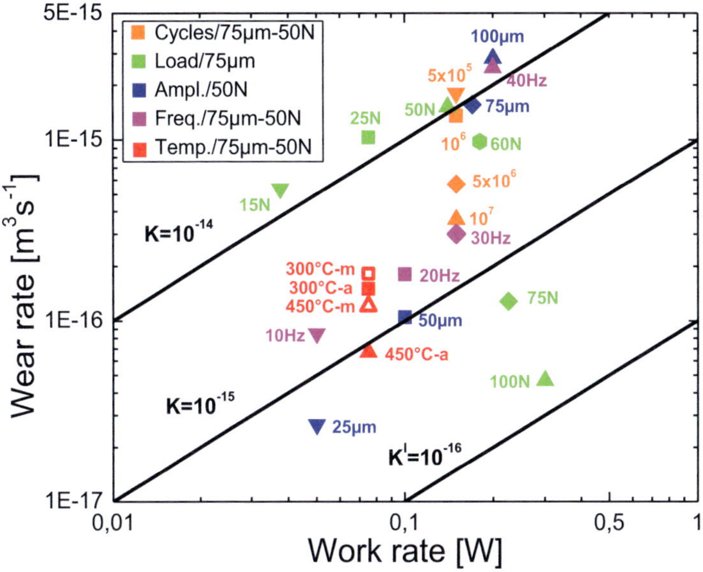

Fig.5.26. Graph wear rate vs work rate for different fretting conditions.

Fig. 5.26 highlights that short amplitudes, high loads and high temperature reduce the fretting wear effect. Fig. 5.26 contains values of wear rate for different fretting conditions (work rates) so that it also offers a tool for predicting fretting wear damage in the time with a certain safety margin. Indeed, most of the experiments were carried out within the run-in period when fretting is generally severe at the most.

6. Fretting tests in molten lead: results and discussion

6.1 Results

6.1.1 General appearance of the fretting damage

The nature of the fretted areas observed on friction pairs tested in liquid lead is schematically shown in Fig. 6.1. The material removal resulting from the fretting action causes the formation of a groove on the original surface and a reduction of the specimens (and counter-specimens) cross-section. Beside, on the fretted areas the presence of not homogeneous and not constant scales, which interposes between specimen and counter specimen during the fretting motion, can be observed. As shown in the simplified sketch of Fig. 6.1 in such compacted scale two different layers, both enriched in Pb, are distinguishable: an outer layer mainly composed of oxidized debris and an inner layer mainly consisting in metallic debris. In some particular cases the outer or the inner layer might be not clearly visible.

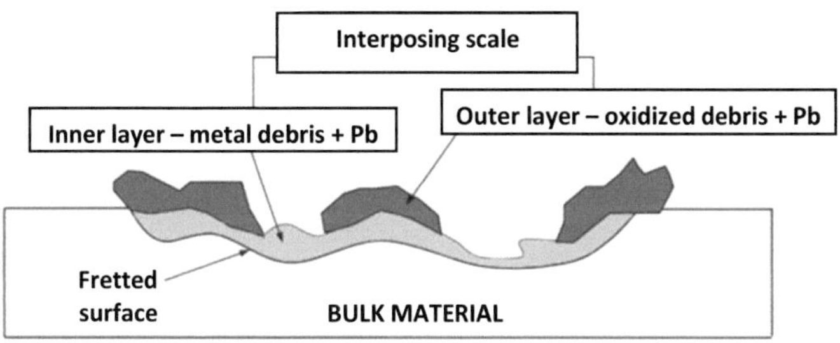

Fig.6.1. Simplified illustration of the fretting affected area. An outer layer made of oxidized debris (dark grey) and an inner made of metal debris (light grey) can be noticed.

Thickness, extension, homogeneity and composition of the compacted interposing scale vary with test conditions and type of material. These statements are supported by SEM and EDX analysis performed on the tested specimens cross-sections. Descriptive examples for the three alloys under investigation are reported hereafter.

Friction pairs of T91

For specimens of T91 steel, the compacted scale usually consists in an outer layer mostly made of Fe-oxides and an inner layer constituted by debris of the original alloy. Accordingly, the SEM image in BSE mode of Fig. 6.2 and the EDX data reported in Tab. 6.1 show an outer layer (points 1-4) mainly composed of magnetite (Fe_3O_4) with traces of Fe-Cr oxides (most probably Fe-Cr spinel oxides) and Pb. In the inner layer, Cr and Fe in concentrations close to the T91 nominal ratio (Cr: Fe = 1:10) and Pb up to ~ 14 wt% (points 5-9) are detected. Points 10 and 11 in Fig. 6.2 and Tab. 6.1 indicate the bulk material unaltered with respect to the nominal composition.

The microstructure analysis of the fretted area of a T91 specimen, accomplished by LOM (Fig. 6.3) and SEM (Fig. 6.4) confirms the formation of a compacted scale. In Fig. 6.3 the grain boundaries of a typical ferritic-matensitic structure (brighter area) characterize the bulk of the T91 specimen. Above the bulk, a compacted scale about 10-15 μm thick without any recognisable ordered structure (dark grey) can be observed. The SEM imaging in BSE mode of Fig. 6.4 agrees with these observations. Here, because of the higher magnification, three main areas can be identified: an outer layer about 5 μm thick of compacted oxidized debris (dark grey band) with some traces of Pb (lighter spots), an inner layer of compacted metal debris (lighter band), enriched in Pb and without organised structure and the bulk with the ferritic-martensitic structure (grey area).

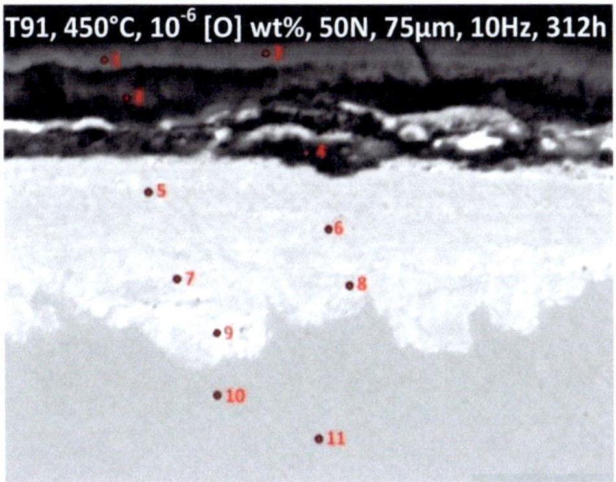

Fig.6.2. SEM image of the typical compacted scale forming on the fretted area of T91 specimens. An outer layer of compacted oxides (dark grey band) and an inner layer of compacted metal debris (light grey band) can be noticed. Test parameters: 450°C, 10^{-6} [O] wt%, 50 N, 75 µm, 10 Hz, 312 h, $1.12 \cdot 10^{7}$ cycles.

(wt%)	O	Cr	Fe	Pb
1	29.6	0.4	66.9	3.0
2	32.3	0.3	63.6	3.6
3	21.3	1.2	76.3	1.0
4	22.1	11.7	52.1	13.9
5	0.0	6.7	80.7	12.5
6	0.0	8.1	83.9	7.8
7	0.0	8.0	81.0	10.8
8	0.0	8.5	82.3	9.1
9	0.0	8.0	81.2	10.6
10	0.0	8.4	89.4	0.1
11	0.0	8.7	89.7	0.2

Tab.6.1. EDX point analysis performed in the points marked by red numbers of Fig. 6.2

Fig.6.3. LOM image of the fretted area cross-section of a T91 specimen. The typical T91 ferritic-martensitic microstructure (lighter area) and the compacted scale on the fretted area (dark grey band) can be noticed. Test parameters: 450°C, 10^{-6} [O] wt%, 50 N, 75 µm, 10 Hz, 312 h, $1.12 \cdot 10^{7}$ cycles.

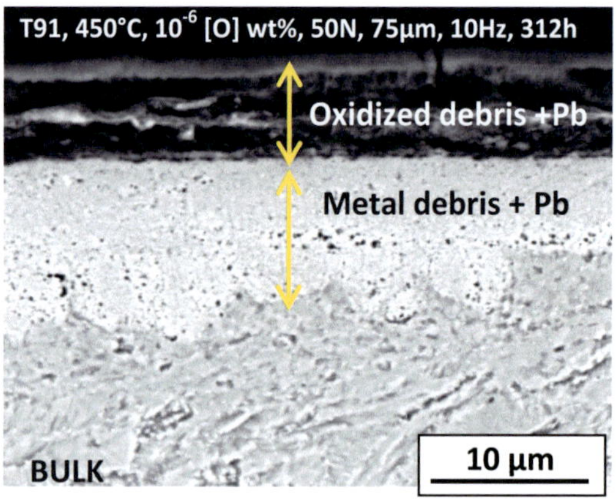

Fig.6.4. SEM image in BSE mode of the fretted area a T91 specimen. Compacted oxidized debris (darker band), compacted metal debris enriched in Pb (lighter band) and the T91 bulk (grey area) can be noticed. Test parameters: 450°C, 10^{-6} [O] wt%, 50 N, 75 µm, 10 Hz, 312 h, $1.12 \cdot 10^{7}$ cycles.

Friction pairs of 15-15Ti

The main features of the typical compacted scale forming on 15-15Ti specimens during the fretting process are highlighted by the SEM image in BSE mode of Fig. 6.5 and the EDX data of Tab. 6.2. An outer layer, 3-4 μm thick, Ni depleted and containing traces of oxides (darker spots - points 1-2) and an inner layer, 5-10 μm thick, Ni depleted, enriched in Pb and with Cr in nominal amount (points 3-8) can be noticed. Directly beneath the compacted scale, in the bulk, the expected nominal composition is detected (points 10-11).

The formation of the compacted scale finds confirmation in the microstructure analysis executed by LOM and SEM (Fig. 6.6 and 6.7 respectively) after etching of the same cross-section displayed in Fig. 6.5.

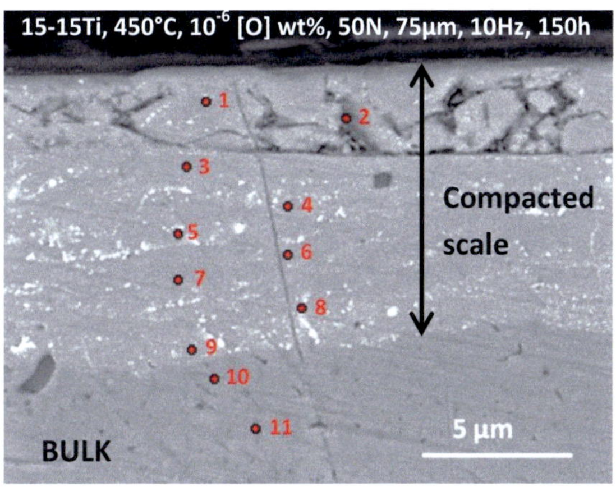

Fig.6.5. SEM image of the typical compacted scale forming on 15-15Ti. An outer layer of oxidized debris (darker spots) and an inner layer of metal debris can be noticed. Test parameters: 450°C, 10^{-6} [O] wt%, 50 N, 75 μm, 10 Hz, 159 h, $5.4 \cdot 10^{6}$ cycles.

(wt%)	O	Cr	Fe	Ni	Pb
1	1.7	13.6	74.6	0.7	9.1
2	1.7	14.0	77.6	0.7	5.8
3	0.0	14.4	77.5	1.2	6.7
4	0.0	14.0	77.8	1.2	6.8
5	0.0	15.1	76.6	1.0	7.2
6	0.0	15.3	77.2	1.7	5.6
7	0.0	14.4	76.0	1.9	7.6
8	0.0	14.6	73.1	7.4	4.7
9	0.0	14.9	66.3	14.3	4.3
10	0.0	15.7	68.9	14.8	0.4
11	0.0	15.4	68.5	15.3	0.6

Tab.6.2 EDX point analysis performed in the points marked by red numbers of Fig. 6.5

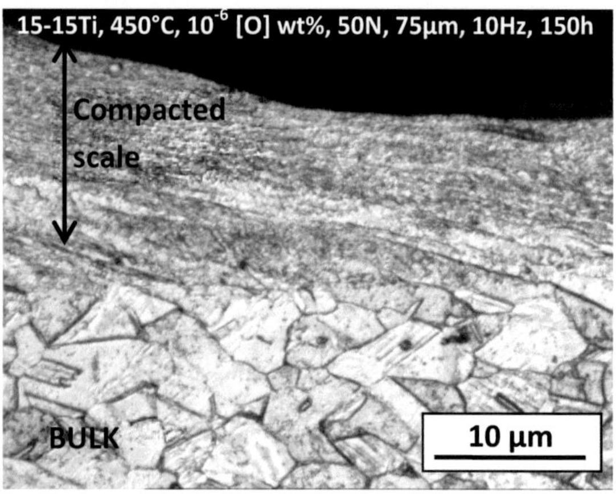

Fig.6.6. LOM image of the fretted area cross-section of a 15-15Ti specimen. Austenitic microstructure and the compacted scale can be noticed. Test parameters: 450°C, 10^{-6} [O] wt%, 50 N, 75 μm, 10 Hz, 150 h, $5.4 \cdot 10^{6}$ cycles.

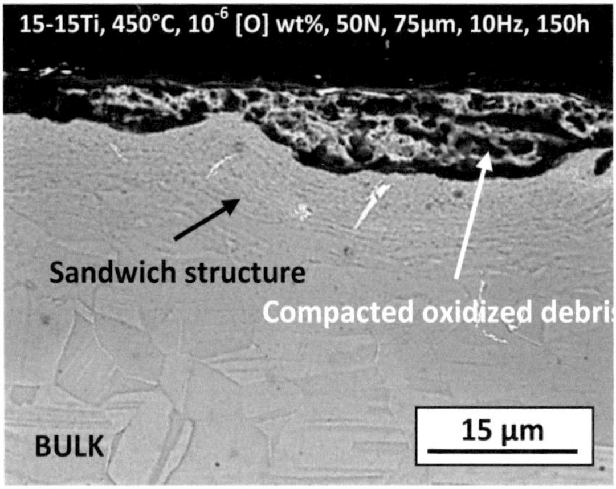

Fig.6.7. SEM image BSE mode of the fretted area of a 15-15Ti specimen. The outer band of compacted oxidized debris (black/light grey), the inner band of compacted deformed metal debris and Pb and the bulk are visible. Test parameters: 450°C, 10^{-6} [O] wt%, 50 N, 75 µm, 10 Hz, 150 h, $5.4 \cdot 10^{6}$ cycles.

In the LOM image (Fig. 6.6), a compacted scale without any specific ordered structure and spread on the underlying typical austenitic structure of the 15-15Ti steel can be observed. The SEM image in BSE mode (Fig. 6.7) suggests that the compacted scale consists more externally of a compound of oxides (darker spots) and Pb (lighter area) and more internally of a kind of sandwich-like structure, in which deformed austenitic grains (fretting debris) are mixed and compacted with Pb. The austenitic structure is visible underneath.

On the 15-15Ti, differently from the other tested alloys, dissolution attack in the fretted area was noticed after tests at specific conditions. As example of a dissolution affected fretted area, in Fig. 6.8 the SEM image in BSE mode and EDX line scan of the fretted area resulting after test a of 930 h at 450°C is shown. In this picture, the presence of a dissolution affected area (B band) directly underneath the compacted scale (A band) can be noticed.

Such area is characterized by Pb penetration into the steel matrix for about 20 µm, almost complete Ni depletion and partially Cr depletion. Underneath the dissolution area bulk material (C band) is visible.
Here, as shown by the EDX lines, Ni and Cr reach the nominal concentration (15 wt% of Cr and 15 wt% of Ni).

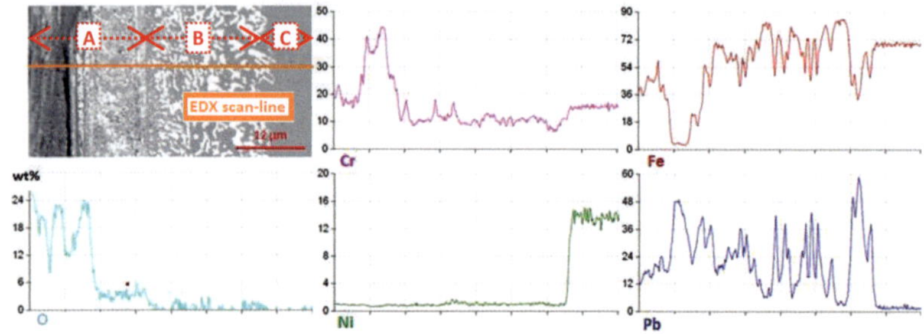

Fig.6.8 SEM image in BSE mode and EDX line analysis (orange line) of a 15-15Ti specimen cross-section after fretting test. Compacted scale (A-band), dissolution affected area (B-band) and bulk material (C-band) can be noticed. Test parameters: 450°C, 10^{-6} [O] wt%, 50 N, 75 µm, 10 Hz, 930 h, $3.35 \cdot 10^{7}$cycles.

Friction pairs of GESA-T91

Regarding GESA-T91 specimens, the surface alloyed layer is gradually removed by the fretting action. The debris forming during the fretting process are compacted on the fretted surface in a similar manner to 15-15Ti and T91 steels. As displayed by the SEM image in BSE mode of Fig. 6.9, in the fretted area, thinning of the alloyed layer and formation of a compacted scale (~10-15 µm thick) can be observed. The EDX data collected in Tab. 6.3, suggest that the outer part of such compacted scale is mostly made of oxidized debris and Pb (points 1-3), whereas the inner part consists in metal debris mixed with Pb (points 4-6). In the compacted scale, localized partial Al depletion occurs (e.g. points 5-6 in Tab. 6.3). Beneath the compacted scale, the remaining part of the GESA alloyed scale (not worn away by fretting) does not show any composition variation (points 7-

8 Tab. 6.3). Points 9 (black spot) and 10-11 in Fig. 6.9 and Tab. 6.3 indicate an Al oxide inclusion (most probably embedded during the substrate cleaning procedure that precedes the powder deposition by LPPS) and the T91 substrate respectively.
The microstructure analysis of a GESA-T91 specimen after a fretting test is reported in Fig. 6.10 and at higher magnification in Fig. 6.11.

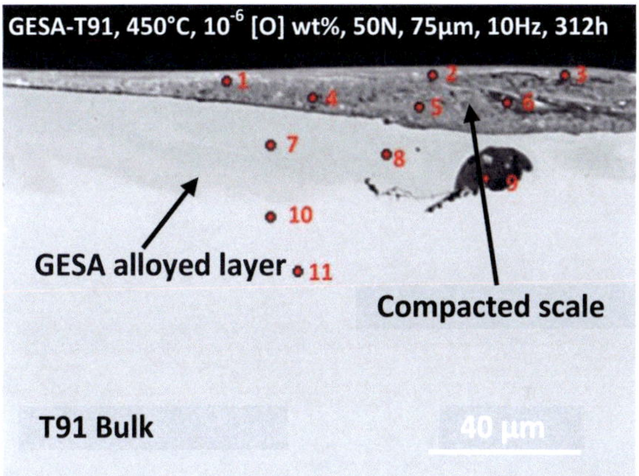

Fig.6.9. SEM image of the compacted scale forming on the fretted area of GESA-T91 (darker grey band). Part of the alloyed layer is visible (grey band) on the T91 substrate (brighter grey part). Test parameters: 450°C, 10^{-6} [O] wt%, 50 N, 75 µm, 10 Hz, 312 h, $1.12 \cdot 10^{7}$ cycles.

(wt%)	O	Al	Cr	Fe	Pb
1	26.32	3.53	7.55	51.37	11.22
2	23.79	3.16	8.86	57.44	6.74
3	27.37	7.52	13.56	43.95	7.60
4	0.00	8.35	12.36	75.66	3.63
5	0.00	3.58	10.46	82.92	3.05
6	0.00	1.18	9.17	87.35	2.31
7	0.00	7.02	11.24	81.04	0.71
8	0.00	8.35	12.36	78.66	0.63
9	37.66	51.72	7.38	1.09	2.15
10	0.00	0.72	8.96	89.87	0.45
11	0.00	0.66	8.94	90.35	0.06

Tab.6.3. EDX point analysis performed in the points marked by red numbers of Fig. 6.9.

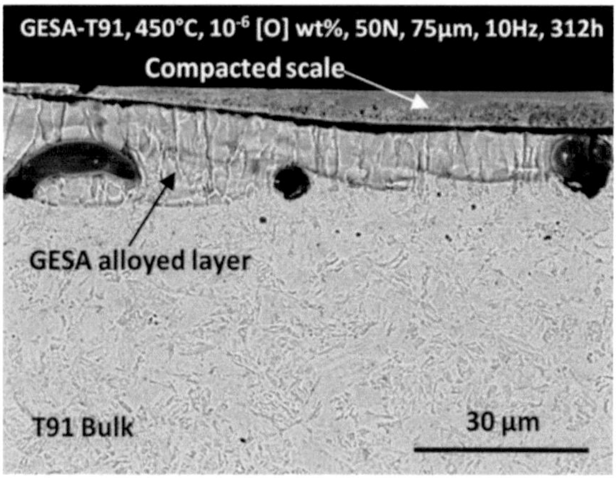

Fig.6.10. SEM image in BSE mode of the fretted area cross-section of a GESA-T91 specimen. Compacted scale and thinning of the GESA scale can be observed. Test parameters: 450°C, 10^{-6} [O] wt%, 50 N, 75 µm, 10 Hz, 312 h, $1.12 \cdot 10^{7}$ cycles.

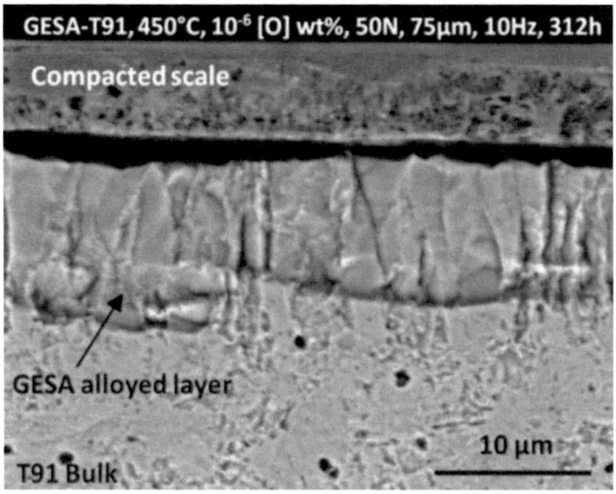

Fig.6.11. SEM image in BSE mode of the fretted area cross-section of a GESA-T91 specimen. Compacted scale and thinned GESA scale are visible. Test parameters: 450°C, 10^{-6} [O] wt%, 50 N, 75 µm, 10 Hz, 312 h, $1.12\cdot10^{7}$ cycles.

In these SEM images in BSE mode, the thinning in the fretted area (up to ~8 µm) of the GESA alloyed surface layer (Fig. 6.10), the compacted scale formed (up to ~8 µm thick) and the T91 bulk beneath are visible. In Fig. 6.11, it is possible to notice a compacted scale without any specific organized microstructure and the typical longitudinally elongated grains characterizing the GESA alloyed layer beneath. Besides, the underlying ferritic –martensitic structure of the T91 bulk can be discerned.

6.1.2 *Influence of exposure time/number of cycles*

The evolution of the fretting wear and corrosion process with the time/number of cycles was investigated by performing tests with fixed values of applied load (50 N), sliding amplitude (75 µm), sliding frequency (10 Hz), temperature (450°C) and oxygen content (10^{-6} wt%). Exposure time and corresponding number of cycles were varied in the range 150-930 h

and $5.4 \cdot 10^6 - 3.35 \cdot 10^7$ respectively. Values of sliding amplitude and applied load were chosen considering the experimental outcomes in air (paragraph 5.1), in order to establish severe fretting conditions.
Volume loss and fretting depth versus number of cycles/time for the friction pairs of T91, 15-15Ti and GESA-T91 steels are reported in Fig. 6.12 a 6.13 respectively. These parameters increase with time (number of cycles) for all the tested materials. GESA-T91 shows the lowest values of volume loss and fretting depth, while 15-15Ti exhibits the highest. For 15-15Ti and T91, the increase rate seems faster during the first 312 h than afterwards. On the contrary, for GESA-T91 the curve slope increases with increasing time.
With reference to Fig. 6.12 and 6.13, after 930 h ($\sim 3.35 \cdot 10^7$ cycles), for 15-15Ti volume loss and fretting depth are 0.23 mm^3 and 111 μm (22.2 % penetration) respectively. Such values for T91 and GESA-T91 are 0.153 mm^3 / 72 μm (14.4 % penetration) and 0.082 mm^3 / 37 μm (7.4 %) respectively. The fretting wear of GESA-T91 is around 1/2 compared to T91 and 1/3 compared to 15-15Ti.
The wear rate, plotted in Fig. 6.14, confirms that for T91 and 15-15Ti the fretting wear is more severe in the early stage of the process whereas for GESA-T91 the converse is true. For T91 and 15-15Ti the wear rate, after a slight increase in the first 312 h ($\sim 1.1 \cdot 10^7$ cycles), decreases down to a minimum value (after 930 h) of $\sim 4.5 \cdot 10^{-8}$ and $\sim 7 \cdot 10^{-8}$ mm^3s^{-1} respectively. For GESA-T91, the wear rate increases from $\sim 1.2 \cdot 10^{-8}$ mm^3s^{-1} (after 312 h) to $\sim 2.5 \cdot 10^{-8}$ mm^3s^{-1} (after 930 h). The specific wear coefficient (Fig. 6.15) follows a trend similar to the wear rate; after 312 h, it decreases for T91 and 15-15Ti and increases for GESA-T91.

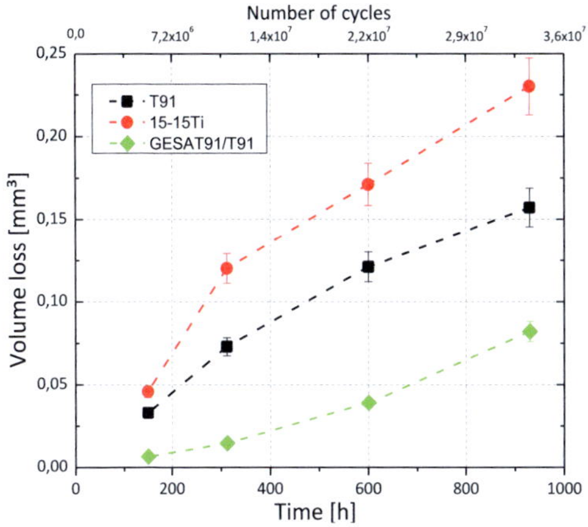

Fig.6.12. Graph of volume loss versus time/number of cycles for friction pairs of GESA-T91 (green diamonds), T91 (black squares) and 15-15Ti (red circles). Test parameters: 75 µm, 50 N, 10 Hz, 450°C, 10^{-6} [O] wt%.

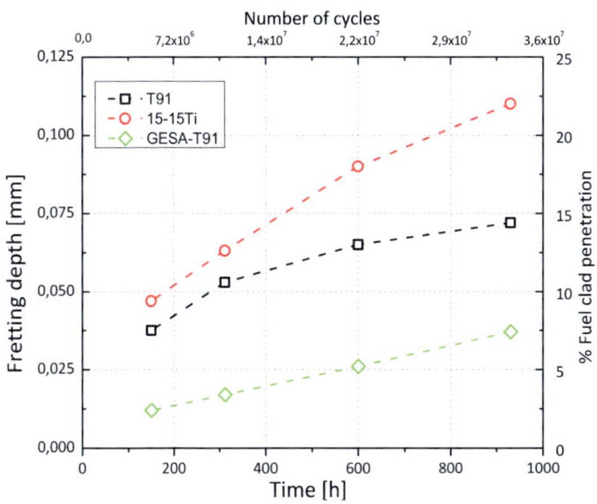

Fig.6.13. Graph of fretting depth versus time/number of cycles for friction pairs of GESA-T91 (green diamonds), T91 (black squares) and 15-15Ti (red circles). Test parameters: 75 µm, 50 N, 10 Hz, 450°C, 10^{-6} [O] wt%

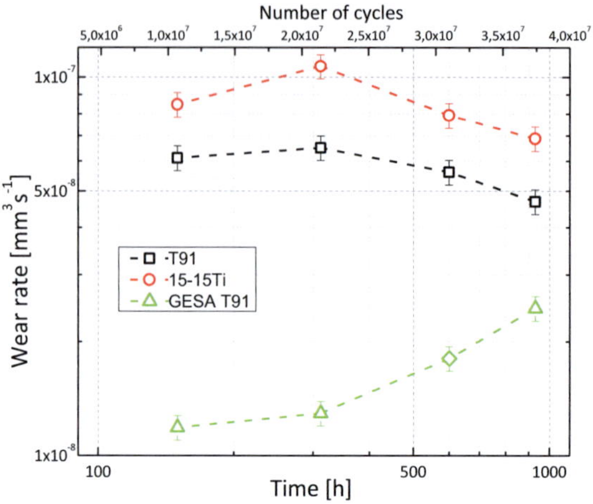

Fig.6.14. Graph of wear rate versus time/number of cycles for friction pairs of GESA-T91 (green diamonds), T91 (black squares) and 15-15Ti (red circles). Test parameters: 75 μm, 50 N, 10 Hz, 450°C, 10^{-6} [O] wt%.

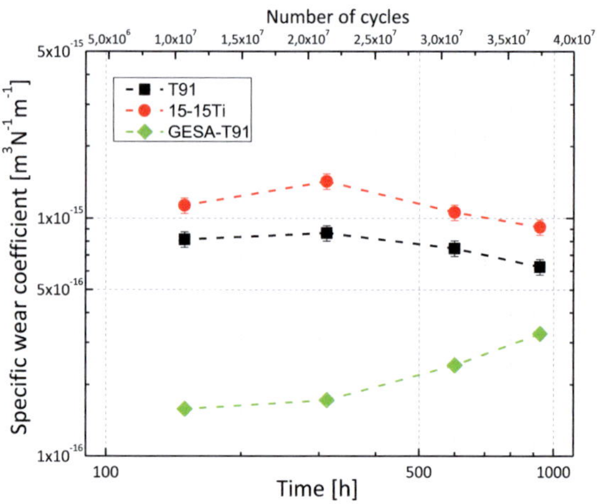

Fig.6.15. Graph of specific wear coefficient versus time/number of cycles for friction pairs of GESA-T91 (green diamonds), T91 (black squares) and 15-15Ti (red circles). Test parameters: 75 μm, 50 N, 10 Hz, 450°C, 10^{-6} [O] wt%.

Metallographic observations

On the area of the specimens (of the tested alloys) not affected by fretting protective oxide scales grow and, for the investigated exposure time and temperature, no dissolution or extensive oxidation occur.

On the not fretted surfaces of T91 (SEM image of Fig. 6.16), a duplex oxide scale made of an outer layer of magnetite (Fe_3O_4 - EDX data points 1-4 of Tab. 6.4) and an inner layer of Fe-Cr oxides (most probably spinel oxide - EDX data points 5-9 of Tab. 6.4) develops. The total thickness of such scale generally increases with the exposure time in liquid lead. At 450°C, the total oxide scale thickness increases from ~4 μm after 150 h to ~12 μm after 930 h.

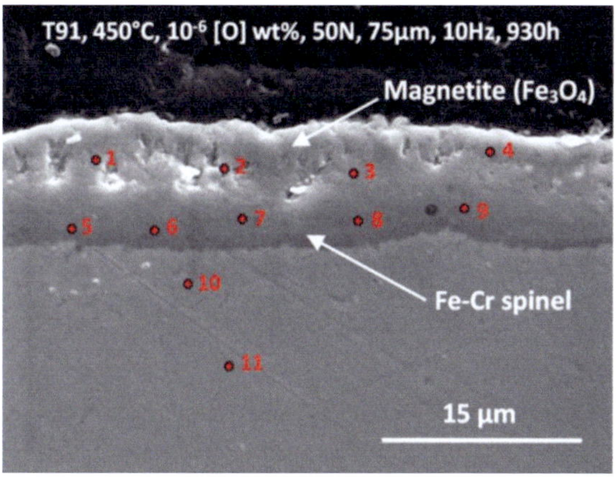

Fig.6.16. SEM image of the duplex oxide scale forming on the not fretted surface of T91 specimens after 930 h fretting test in lead with 10^{-6} wt% of oxygen at 450°C.

(wt%)	O	Cr	Fe	Pb
1	29.1	0.2	65.9	4.7
2	28.7	0.4	63.4	7.4
3	28.8	1.9	62.1	7.1
4	28.4	0.5	64.5	6.4
5	28.6	12.3	56.7	2.2
6	28.0	13.3	57.1	1.3
7	28.6	15.5	54.7	1.1
8	28.1	14.9	55.8	1.0
9	23.3	15.2	59.4	1.9
10	0.4	8.9	89.8	0.7
11	0.9	8.8	89.6	0.3

Tab.6.4. EDX point analysis performed in the points marked by red numbers of Fig. 6.16.

According to the SEM/EDX analysis of Fig. 6.17 and Tab. 6.5 (points from 1 to 9), after exposure in molten Pb with 10^{-6} wt% of oxygen at 450°C, the friction pairs of 15-15Ti are covered (in the not fretted areas) by Fe-Cr oxides scale (most probably spinel type). Such scale is thinner and less homogeneous than the duplex oxide scale noticed on T91 specimens. Indeed, in some areas of the steel surface no constant scale but oxide islands were noticed (Fig. 6.17). The oxide scale (or islands) thickness increases from 2 to 4 μm between 150 and 930 h.
On GESA-T91 specimens, due to the Al containing surface alloying, an alumina (Al_2O_3) oxide scale most probably develops. Unfortunately, at this temperature (450°C) and for these exposure times, the nano-metric scale is hardly detectable with the characterization methods used in this work.
The fretted areas are characterized by removal of the protective oxide scale and formation of a compacted interposing scale (as described in paragraph

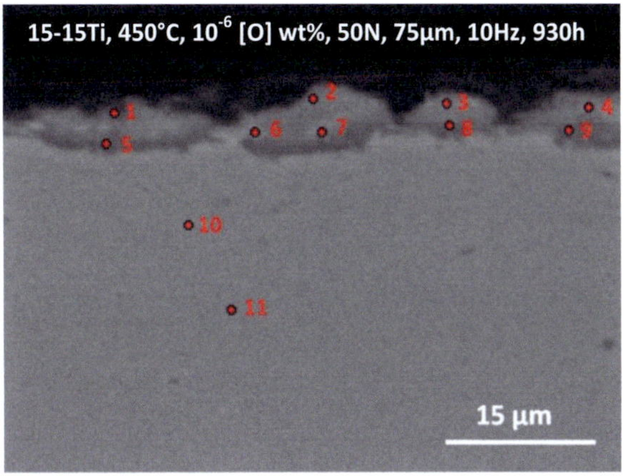

Fig.6.17 SEM image of the typical islands of Fe-Cr oxides forming on the not-fretted surface of 15-15Ti after 930 h test in lead with 10^{-6} wt% of oxygen at 450°C.

(wt%)	O	Cr	Fe	Ni	Pb
1	36.3	2.0	55.1	1.1	5.3
2	34.3	3.7	53.2	1.1	7.4
3	28.5	5.1	57.8	2.7	5.7
4	28.1	13.0	49.0	5.1	4.6
5	26.0	11.7	53.5	4.6	4.0
6	27.2	13.7	52.9	3.2	2.7
7	27.7	14.6	48.9	4.3	4.2
8	29.6	13.8	52.7	2.0	1.7
9	27.9	14.2	55.4	1.2	1.0
10	0.0	15.5	70.0	14.9	0.4
11	0.0	15.6	68.5	15.2	0.7

Tab.6.5. EDX point analysis performed in the points marked by red numbers of Fig. 6.17.

6.1.1). As illustrated in Fig. 6.18 for the fretted area of a T91 specimen, the compacted scale (lighter band) is up to 10-20 μm thick (like in most of the cases) and it is spread on the majority of the fretting groove. Its thickness, regardless of the type of material, is not remarkably influenced by the exposure time.

In the case of friction pairs of 15-15Ti, submitted to fretting test of 312, 600 and 930 h, dissolution attacks were detected in the fretted area. After 312 h, dissolution attacks less than 5 μm deep were noticed in few small and localized areas (beneath the compacted scale covering the fretted area). The extent of dissolution increases with time so that after the 930 h test dissolution is homogeneously distributed along all the fretted area and it is up to 20-25 μm deep. As shown in Fig. 6.8 and 6.19, for a 15-15Ti specimen after 930 h exposure in liquid lead at 450°C, the dissolution attack starts beneath the compacted scale and spreads inwards. In line with these observations, after 600 h, dissolution attacks are more localized than after 930 h and they are about 10 μm deep. Additionally, after the 600 h fretting test, some cracks starting from the fretting area and propagating into the bulk material were detected. As shown by the SEM image and EDX Ni mapping of Fig. 6.20, these cracks are up to about 60 μm long and favour deep lead penetration, dissolution attack and consequently Ni depletion.

The specimens of GESA-T91, as mentioned in paragraph 6.1.1, due to the fretting action are subjected to the progressive removal of the GESA alloyed surface layer. As visible in the SEM image in BSE mode of Fig. 6.21, even after 930 h test and ~$3.35 \cdot 10^6$ fretting cycles, there are few places where rests of the GESA treated layer are still visible.

The upper micrograph of Fig. 6.21 shows the border between the fretted area, where the GESA alloyed layer is removed, and the not fretted area, where the original GESA alloyed layer is unaffected. The lower micrograph of Fig. 6.21 shows a portion of the fretting area where a residual part of GESA alloyed layer (~ 5 μm thick) is still visible beneath a compacted scale of about 10 μm.

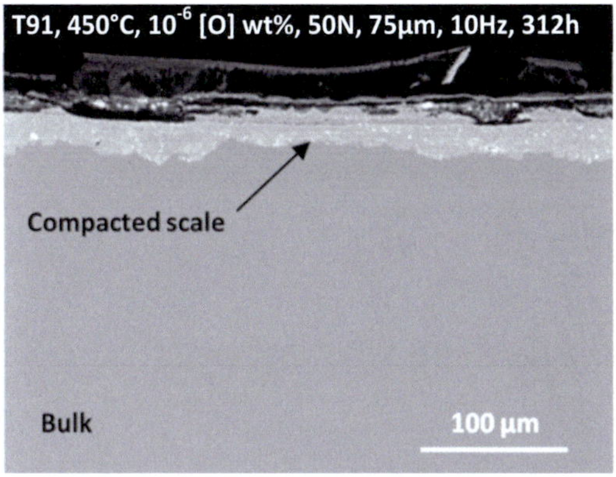

Fig.6.18 SEM image in BSE mode of the fretted area of T91. The lighter band is the compacted scale forming during fretting.

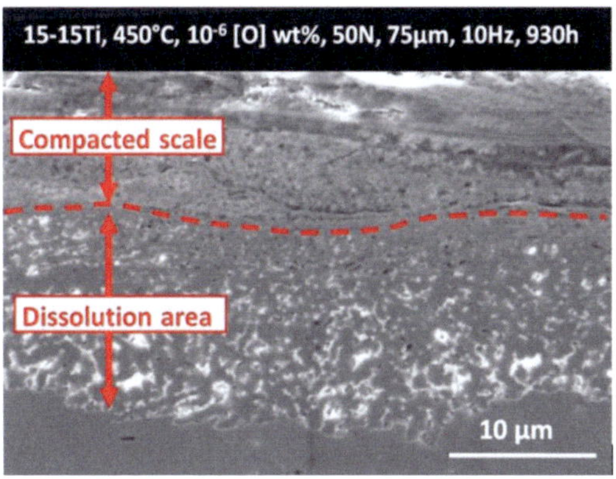

Fig.6.19 SEM image in BSE mode of 15-15Ti. Under the compacted scale, Pb penetrates the bulk (brighter spots) and dissolution occurs.

Fig.6.20 SEM image and EDX Ni mapping of the fretting area cross-section of a 15-15Ti specimen. The SEM picture (left) shows cracks (dark grey) propagating in the bulk material and favors lead penetration (lighter spot). The EDX Ni mapping (right) confirms Ni depletion (black area) in the dissolution area.

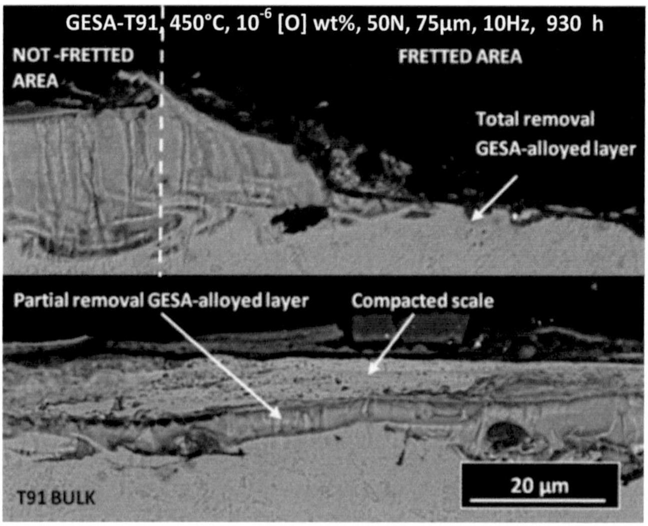

Fig.6.21. SEM imaging in BSE mode executed in two different regions of the fretted area resulting on a friction pair of GESA-T91 after 930 h test.

6.1.3 Influence of applied load

The influence of the applied load on the fretting process was investigated performing tests in molten Pb with a fixed exposure time/ number of cycles (150 h / 5.4·10^6 cycles), sliding amplitude of 35, 75 and 165 µm, sliding frequency (10 Hz), temperature (450°C) and oxygen content (10^{-6} wt%). Different loads were applied in the range between 15 and 75 N. The f/m T91 steel was the material selected for these tests. Besides, for comparison purposes, tests on 15-15Ti were also performed.

As shown in Fig. 6.22 and 6.23, for friction pairs of T91, the fretting wear (volume loss and fretting depth) increases with the increasing applied load up to a maximum (turning point) and then it decreases. The value of the applied load corresponding to the turning point varies in dependence of the imposed amplitude.

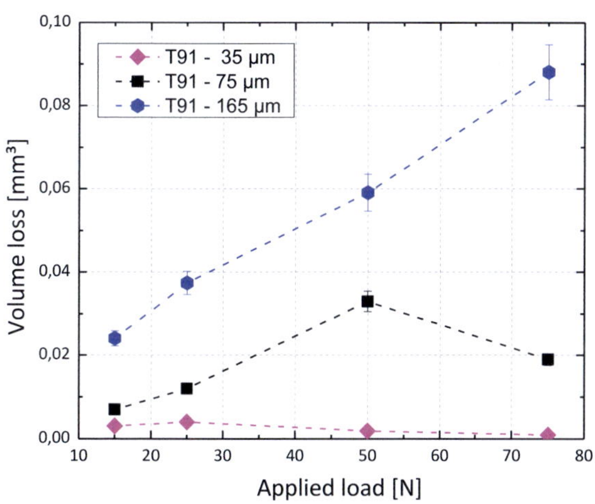

Fig.6.22. Graph of volume loss vs applied load for T91. Test parameters: 35 µm, 75 and 165 µm, 10 Hz, 150 h, 5.4·10^6 cycles, 450°C, 10^{-6} [O] wt%.

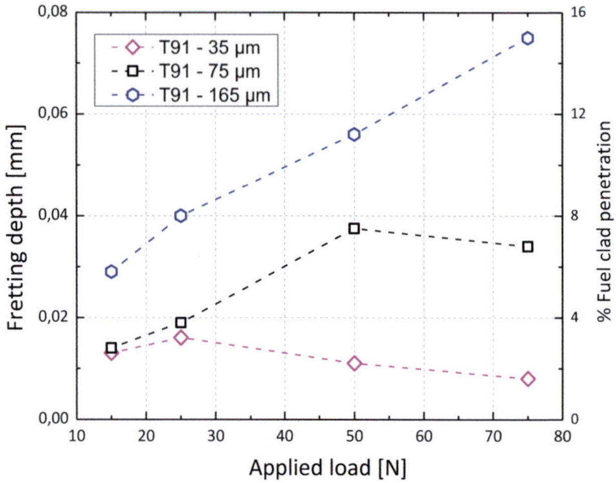

Fig.6.23. Graph of fretting depth vs applied load for T91. Test parameters: 35 µm, 75 µm and 165 µm, 10 Hz, 150 h, $5.4 \cdot 10^6$ cycles, 450°C, 10^{-6} [O] wt%.

With an imposed sliding amplitude of 35 µm, the turning point is placed at 25 N, where volume loss and fretting depth are 0.004 mm^3 and 16 µm (3.2 % penetration) respectively. With 75 µm sliding amplitude, the maximum fretting wear (turning point), corresponding to a volume loss of 0.033 mm^3 and the fretting depth of 37.5 µm (7.4 % penetration), is achieved for 50 N. With imposed sliding amplitude of 165 µm, no turning points is visible (probably exists for higher loads); the data monotonically increase and with an applied load of 75 N the maximum values of 0.088 mm^3 and 75 µm (15 % penetration) are recorded for volume loss and fretting depth respectively. The trend noticed for volume loss and fretting depth is also visible for the wear rate in Fig. 6.24. For example, with imposed sliding amplitude of 35 µm, the wear rate between the turning point (maximum) and the lowest point (75 N) decreases by a factor 7 (from $7.4 \cdot 10^{-9}$ to $1.6 \cdot 10^{-9}$ $mm^3 s^{-1}$).

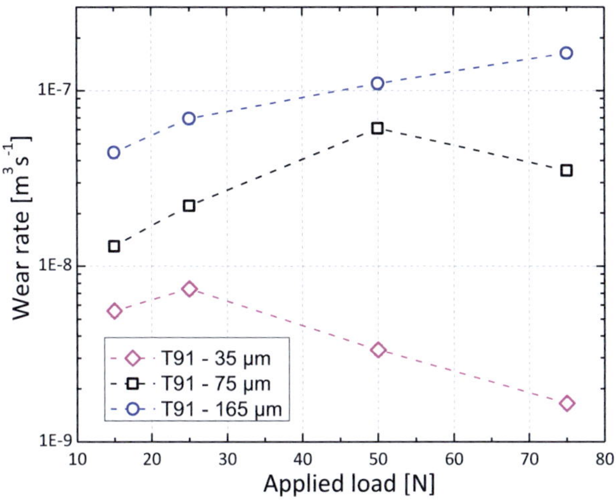

Fig.6.24. Graph of wear rate versus applied loads for friction pairs of T91. Test parameters: 35 µm, 75 µm and 165 µm, 10 Hz, 150 h, $5.4{\cdot}10^6$ cycles, 450°C, 10^{-6} [O] wt%.

Regarding the specific wear coefficient, as shown by Fig. 6.25, this parameter varies only slightly before the turning point and then it decreases remarkably with the increasing load. Indeed both for imposed amplitudes of 35 (pink diamonds) and 75 µm (black squares), before the turning point the maximum variation of specific wear coefficient is around 20-30%. On the contrary, between the turning point and 75 N, the wear rate decreases by a factor 2.5 and by almost an order of magnitude for an imposed amplitude of 75 µm and 35 µm respectively. For an imposed amplitude of 165 µm (blue circles), only small variations of the specific wear rate were measured for increasing load.

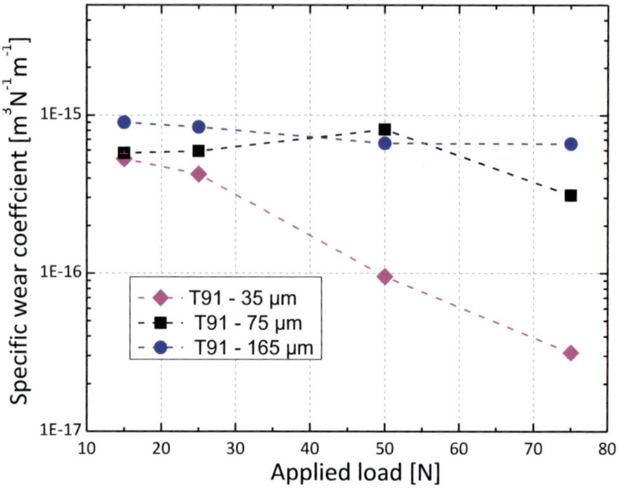

Fig.6.25. Graph of specific wear coefficient versus applied load for friction pairs of T91. Test parameters: 35 µm 75 µm, and 165 µm, 10 Hz, 150 h, $5.4 \cdot 10^6$ cycles, 450°C, 10^{-6} [O] wt%.

To investigate the influence of the applied load on fretting wear of the austenitic steel 15-15Ti, tests with fixed imposed amplitude of 75 µm and loads between 15 and 75 N were performed. To compare the results of these tests with the ones collected for friction pairs of T91, the same testing parameters were adopted.

For the two alloys, the trend of volume loss (Fig. 6.26) and fretting depth (Fig. 6.27) are very similar. The main difference is that the values of volume loss and fretting depth measured on 15-15Ti (red circles) are higher than the ones reported for T91 (black squares). For 15-15Ti, the volume loss is between 30 to 75 % higher than for T91. Additionally, fretting depths measured on 15-15Ti specimens are 20 to 30 % larger than the ones detected on T91 specimens.

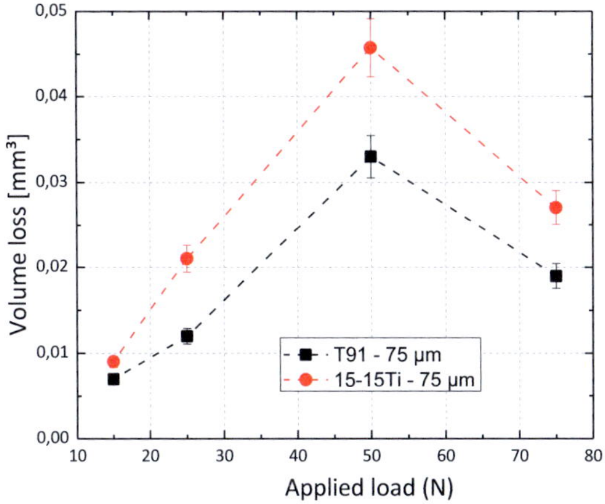

Fig.6.26. Graph of volume loss vs applied load for friction pairs of T91 and 15-15Ti. Test parameters: 75 μm, 10 Hz, 150 h, $5.4 \cdot 10^6$ cycles, 450°C, 10^{-6} [O] wt%.

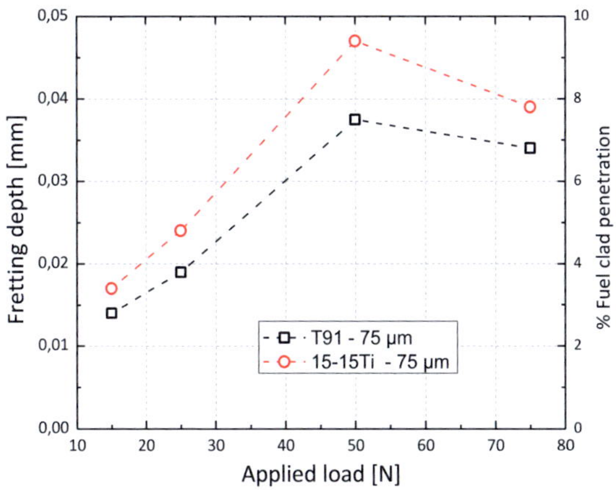

Fig.6.27. Graph of fretting depth vs applied load for friction pairs of T91 and 15-15Ti. Test parameters: 75 μm, 10 Hz, 150 h, $5.4 \cdot 10^6$ cycles, 450°C, 10^{-6} [O] wt%.

The maximum fretting depth (measured at 50 N) for 15-15Ti and T91 is 47 µm (9.4 % penetration) and 37.5 µm (7.5 %) respectively.

The comparisons between wear rates and specific wear coefficients of T91 and 15-15Ti are shown in Fig. 6.28 and 6.29 respectively. For both materials, the wear rate increases up to a turning point corresponding to an applied load of 50 N and then it decreases. The maximum wear rate measured is ~$8.5 \cdot 10^{-8}$ mm^3s^{-1} for 15-15Ti and ~$6 \cdot 10^{-8}$ mm^3s^{-1} for T91.

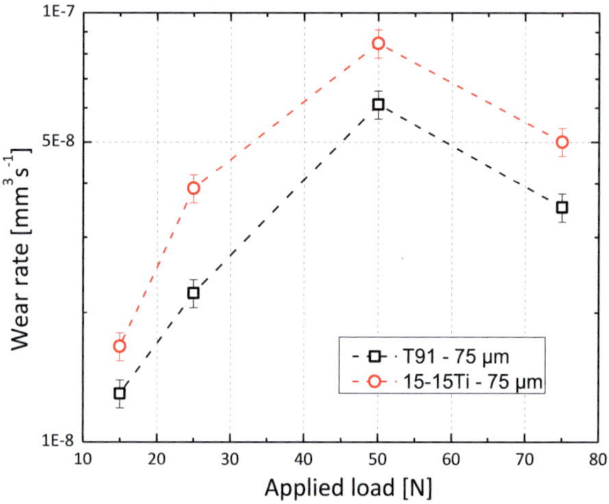

Fig.6.28. Graph of wear rate versus applied load for friction pairs of T91 and 15-15Ti. Test parameters: 75 µm, 10 Hz, 150 h, $5.4 \cdot 10^6$ cycles, 450°C, 10^{-6} [O] wt%.

For both the alloys, the specific wear coefficient (Fig. 6.29) shows a significant decrease for loads higher than 50 N. For 15-15Ti and T91, with an applied load of 75 N, the specific wear coefficient is about 65 % lower than with 50 N.

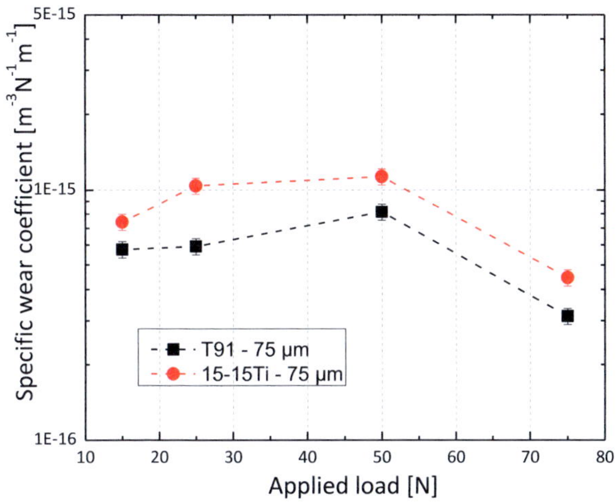

Fig.6.29. Graph of specific wear coefficient versus load for T91 and 15-15Ti. Test parameters: 75 µm, 10 Hz, 150 h, $5.4{\cdot}10^{6}$ cycles, 450°C, 10^{-6} [O] wt%.

6.1.4 *Influence of sliding amplitude*

The role of the sliding amplitude in the fretting process was studied performing tests in molten Pb on friction pairs of T91 with a fixed exposure time/ number of cycles (150 h / $5.4{\cdot}10^{6}$ cycles), 4 different applied loads (15, 25, 50 and 75 N), sliding frequency 10 Hz, temperature 450°C and oxygen content 10^{-6} wt%. Different sliding amplitudes were imposed in the range from 15 to 190 µm.

Volume loss and fretting depth, plotted in Fig. 6.30 and 6.31 respectively, increase with increasing amplitude for all the fixed applied loads.

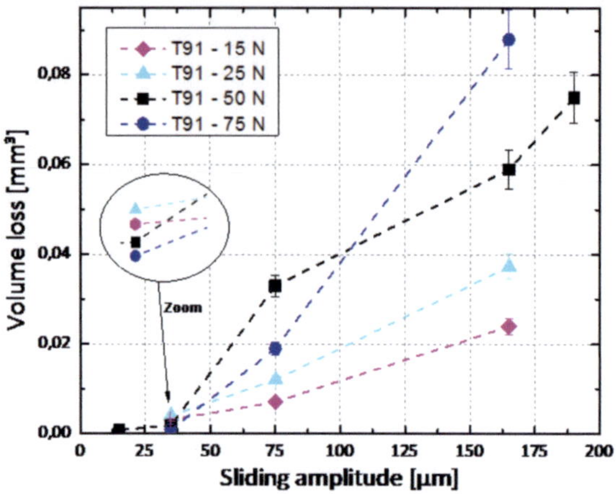

Fig.6.30. Graph of volume loss versus sliding amplitude for friction pairs of T91. Test parameters: 15 N, 25 N, 50 N and 75, 10 Hz, 150 h, $5.4 \cdot 10^6$ cycles, 450°C, 10^{-6} [O] wt%.

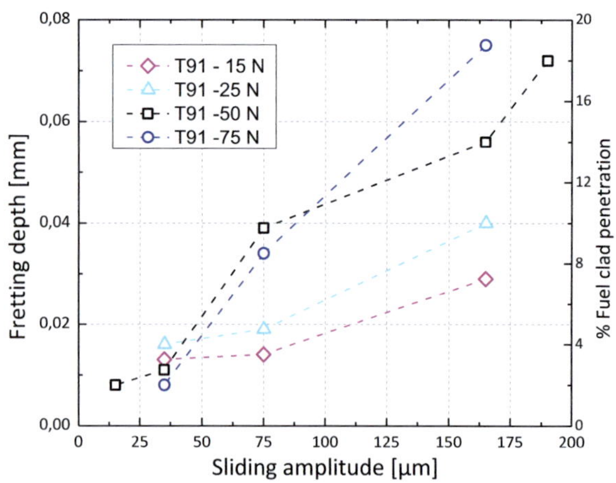

Fig.6.31. Graph of fretting depth versus sliding amplitude for friction pairs of T91. Test parameters: 15 N, 25 N, 50 N and 75 N, 10 Hz, 150 h, $5.4 \cdot 10^6$ cycles, 450°C, 10^{-6} [O] wt%.

After tests of 150 h with sliding amplitude of 165 µm, the fretting depth (Fig. 6.31) reaches 29 µm (5,8 % penetration) with a fixed applied load of 15 N, 40 µm (8 %) with load of 25 N, 56 µm (11.2 %) with 50 N and 75 µm (15 %) with an applied load of 75 N. With an applied load of 50 N, additional tests with sliding amplitude of 15 and 190 µm were performed. Such tests pointed out that volume loss and fretting depth experience a further increase (+ ~20 %) for an amplitude larger than 165 µm and a slight decrease for an amplitude shorter than 35 µm.

In Fig. 6.30 and 6.31, the non-monotonic dependency of fretting wear on load can be noticed, with reference to Fig. 6.22 and 6.23. For a fixed value of amplitude, the highest load not always corresponds to the biggest volume loss and deepest fretting penetration. As put in evidence by the zoomed area of Fig. 6.30, with fixed amplitude of 35 µm, the maximum and minimum volume loss were measured with an applied load of 25 N and 75 N respectively and the same is observable for the fretting depth (Fig.6.31). With sliding amplitude of 75 µm, the biggest damage is caused by an applied load of 50 N and the smallest with an applied load of 15 N. With amplitude of 165 µm, the highest values of volume loss and fretting depth were recorded with an applied load of 75 N and the lowest values with a load of 15 N.

In Fig. 6.32 and 6.33, the wear rate and the specific wear coefficient confirm the observation made so far. A general increase of these parameters with increasing amplitude can be noticed independently from the load applied during the fretting test. However, for a fixed value of sliding amplitude, the higher load does not correspond always to the higher values of wear rate and specific wear coefficient. Maximum and minimum values of wear rate and specific wear coefficient for a fixed value of amplitude follow the same distribution described for volume loss and fretting depth (Fig. 6.30 and 6.31).

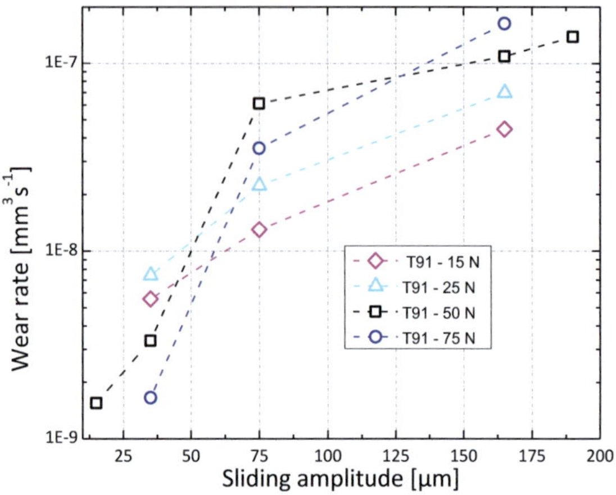

Fig.6.32. Graph of wear rate versus sliding amplitude for friction pairs of T91. Test parameters: 15 N, 25 N, 50 N and 75 N, 10 Hz, 150 h, $5.4 \cdot 10^6$ cycles, 450°C, 10^{-6} [O] wt%.

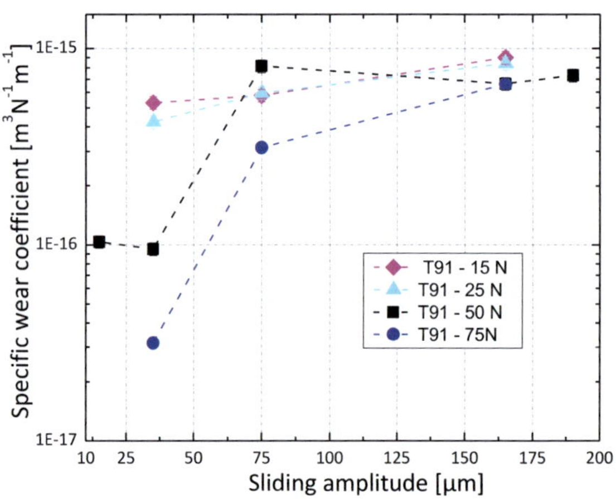

Fig.6.33. Graph of the specific wear coefficient versus sliding amplitude for friction pairs of T91. Test parameters: 15 N, 25 N, 50 N and 75 N, 10 Hz, 150 h, $5.4 \cdot 10^6$ cycles, 450°C, 10^{-6} [O] wt%.

The change of wear rate (Fig. 6.32) and specific wear coefficient (Fig. 6.33) with the increasing amplitude is more moderate for low applied loads (i.e. 15 and 25 N), than for high loads (i.e. 50 and 75 N). Increasing the amplitude from 35 to 165 μm, with applied loads of 15 N and 25 N the wear rate increases by about 1 order of magnitude and the wear coefficient doubles. In the same range of amplitude, with applied loads of 50 and 75 N the wear rate increases by about 2 orders of magnitude and the specific wear coefficient increases of around 7 (with 50 N) and 17 times (with 75 N).

For the specific case of an applied load of 50 N (black squares), between sliding amplitudes of 15 and 35 μm, the specific wear coefficient (Fig. 6.33) is nearly constant and it slightly varies for amplitudes larger than 7μm.

The effect of the amplitude on the fretting process was evaluated also for the austenitic steel 15-15Ti and for the GESA treated T91 steel and compared to the results achieved for the original T91. For this purpose, fretting tests of 150 h with constant applied load of 50 N and amplitude variable between 15 and 190 μm were performed with the three materials under investigation.

The resulting fretting wear is depicted in Fig. 6.34 and 6.35 in terms of volume loss and fretting depth respectively. For the same testing conditions, GESA-T91 shows the smallest damage whereas 15-15Ti the biggest. With shorter sliding amplitude (i.e. 35 μm), volume loss and fretting depth of GESA-T91 (0.007 mm^3 and 10 μm) are very similar to the one of unmodified T91 (0.0011 mm^3 and 11 μm). However, for larger amplitude the difference increases significantly and for a sliding amplitude of 165 μm, volume loss and fretting depth for GESA-T91 (0.031 mm^3 and 32 μm) are about half of the ones of T91 (0.059 mm^3 and 56 μm).

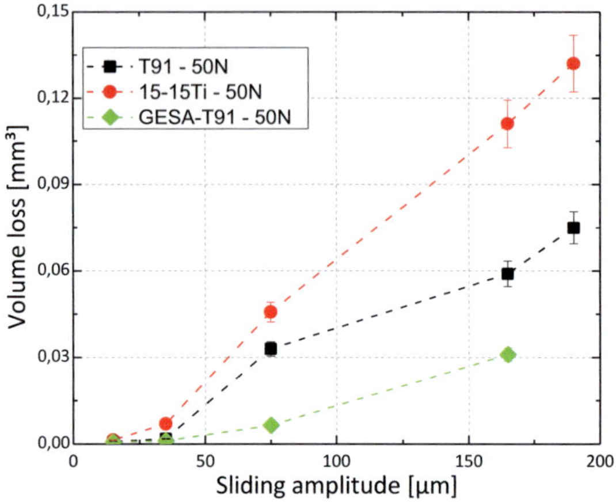

Fig.6.34. Graph of volume loss versus sliding amplitude for friction pairs of T91 (black squares), 15-15Ti (red circles) and GESA-T91 (green diamonds). Test parameters: 50 N, 10 Hz, 150 h, $5.4 \cdot 10^6$ cycles, 450°C, 10^{-6} [O] wt%.

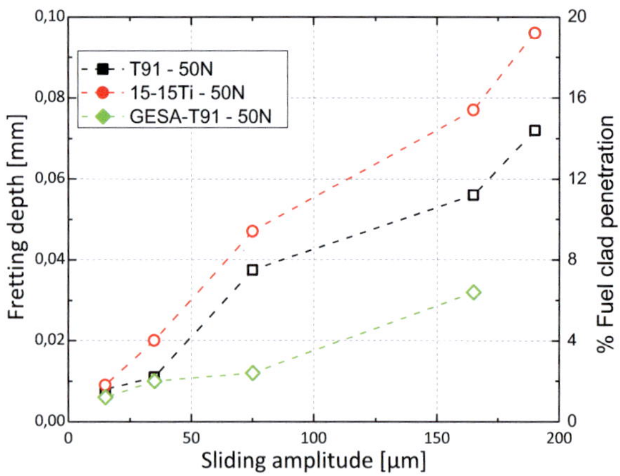

Fig.6.35. Graph of fretting depth versus sliding amplitude for friction pairs of T91 (black squares), 15-15Ti (red circles) and GESA-T91 (green diamonds). Test parameters: 50 N, 10 Hz, 150 h, $5.4 \cdot 10^6$ cycles, 450°C, 10^{-6} [O] wt%.

Friction pairs of 15-15Ti are characterized by the highest values of fretting depth and volume loss for all the tested amplitudes. In particular with a sliding amplitude of 190 µm volume loss/fretting depth for 15-15Ti and T91 are 0.132 mm³/96 µm (19.2 % penetration) and 0.075 mm³/72 µm (14.4 % penetration) respectively.

As shown by the wear rate graph of Fig. 6.36, GESA-T91 is characterized, at all the imposed sliding amplitudes, by the lowest wear rate. On the contrary, 15-15Ti shows the highest values of wear rate, up to an order of magnitude higher than the one of GESA-T91.

For T91 and 15-15Ti, between 15 and 165 µm of amplitude, the wear rate increases by about 2 orders of magnitude. The wear rate of T91 is between $\sim 1\cdot 10^{-9}$ and $\sim 1\cdot 10^{-7}$ $mm^3 s^{-1}$. The one of 15-15Ti, increases from $\sim 2\cdot 10^{-9}$ to $\sim 2\cdot 10^{-7} mm^3 s^{-1}$. In the case of GESA-T91, the wear rate increases less than 2 orders of magnitude; from $\sim 1\cdot 10^{-9}$ to $\sim 6\cdot 10^{-8}$ $mm^3 s^{-1}$.

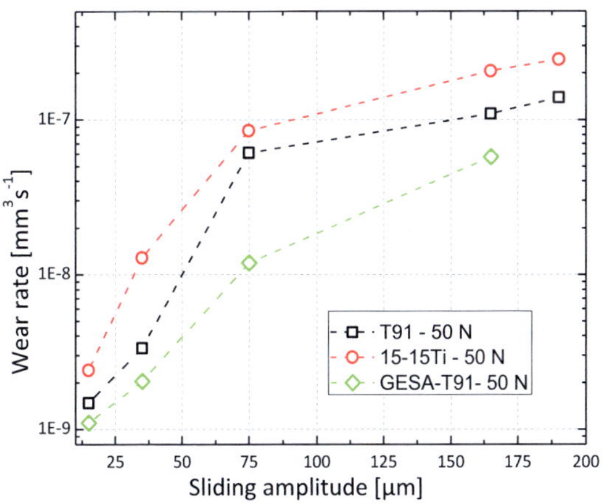

Fig.6.36. Graph of wear rate versus sliding amplitude for friction pairs of T91 (black squares), 15-15Ti (red circles) and GESA-T91 (green diamonds). Test parameters: 50 N, 10 Hz, 150 h, $5.4\cdot 10^{6}$ cycles, 450°C, 10^{-6} [O] wt%.

The specific wear coefficient (Fig. 6.37) of T91 and 15-15Ti increases significantly (> factor 5) between 35 and 75 µm. For both the materials, with sliding amplitude larger than 75 µm (i.e. 165 and 190 µm) and shorter than 35 µm (i.e. 35 µm) the variations of the specific wear coefficient are less evident (< factor 2). For the GESA-T91 the specific wear coefficient increases by about an order of magnitude (from $6{\cdot}10^{-17}$ to $4{\cdot}10^{-16}\,m^3\,N^{-1}m^{-1}$) from 35 to 165 µm, whereas from 15 to 35 µm the variation is more moderate (decreases by a factor 2).

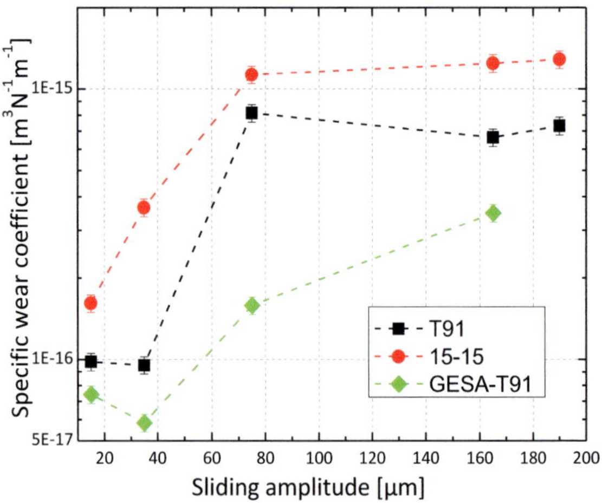

Fig.6.37. Graph of the specific wear coefficient versus sliding amplitude for friction pairs of T91 (black squares), 15-15Ti (red circles) and GESA-T91 (green diamonds). Parameters: 50 N, 10 Hz, 150 h, $5.4{\cdot}10^6$ cycles, 450°C, 10^{-6} [O] wt%.

Metallographic observations

Changes in sliding amplitude influence also the compacted scale formation. The compacted scale is generally thinner and less constant at short sliding amplitudes (e.g. 15 or 35 µm) than at larger amplitudes (e.g. 165 µm).

Besides, the inner part of the compacted scale, the one consisting of metal debris, disappears at short amplitude. These characteristics are shown in SEM micrographs in BSE mode of Fig. 6.38 (with Tab.6.6) and 6.39 (with Tab. 6.7), where the compacted scales detected on T91 specimens after fretting tests with sliding amplitude of 35 and 165 μm are reported respectively. The EDX analysis data (Tab. 6.6), at points marked by red numbers in Fig. 6.38, support the fact that the thin compacted scale (~ 5 μm thick) detected on the specimen tested with sliding amplitude of 35 μm, is mainly composed of oxidized debris.

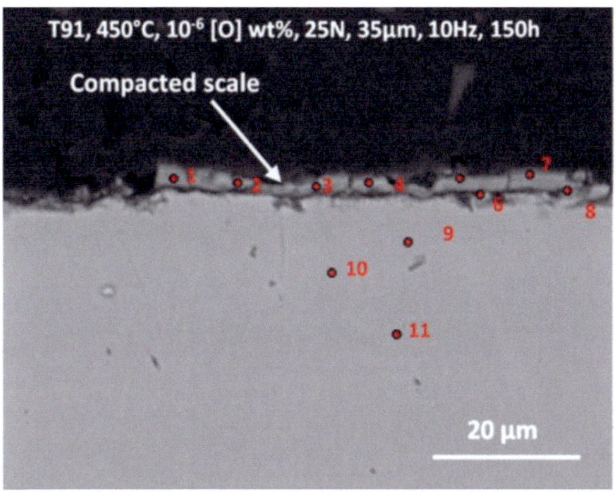

Fig.6.38. SEM image in BSE mode of the compacted oxidized scale detected on T91 after tested with sliding amplitude of 35 μm. Test parameters: 25 N, 10 Hz, 150 h, $5.4 \cdot 10^6$ cycles, 450°C, 10^{-6} [O] wt%.

(wt%)	O	Cr	Fe	Pb
1	20.4	10.9	64.9	3.7
2	24.7	4.6	68.1	2.4
3	20.5	8.5	66.7	4.1
4	30.3	10.1	56.1	3.3
5	22.0	4.7	64.5	8.6
6	15.7	13.6	66.9	3.8
7	26.0	12.5	57.9	3.3
8	15.6	10.1	70.8	3.2
9	0.0	9.7	89.9	0.9
10	0.0	9.1	90.6	0.7
11	0.0	9.2	90.5	0.2

Tab.6.6. EDX point analysis performed in the points marked by red numbers of Fig. 6.38.

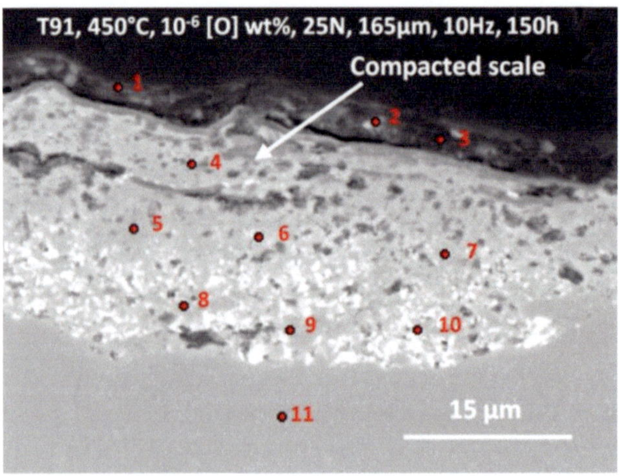

Fig.6.39. SEM image in BSE mode of compacted scale detected on T91 after test with sliding amplitude of 165 µm. Test parameters: 25 N, 10 Hz, 150 h, $5.4 \cdot 10^{6}$ cycles, 450°C, 10^{-6} [O] wt%.

On the contrary, the EDX analysis data (Tab. 6.7), at points marked by red numbers in Fig. 6.39, support the fact that the compacted scale (~ 20 μm thick) detected on the specimen tested with sliding amplitude of 165 μm, consists of an outer layer of oxidized debris (~ 5 μm thick – points 1-3 in Tab. 6.7) and an inner layer of metal debris, both mixed with Pb (~ 15 μm thick– points 4-10 in Tab. 6.7).

(wt%)	O	Cr	Fe	Pb
1	19.4	11.4	65.3	3.7
2	24.1	4.9	68.4	2.4
3	19.0	8.9	67.8	4.1
4	2.8	5.0	83.6	8.4
5	1.1	7.2	82.3	9.2
6	4.9	8.3	78.5	8.1
7	1.1	3.9	87.0	7.7
8	2.8	10.9	63.8	22.3
9	4.1	5.0	72.4	18.3
10	2.2	6.1	63.2	22.4
11	0.0	9.5	89.5	0.7

Tab.6.7. EDX point analysis performed in the points marked by red numbers of Fig. 6.60.

6.1.5 Influence of temperature

The effect of the temperature on the fretting process was investigated performing tests, with a fixed exposure time/ number of cycles (150 h / $5.4 \cdot 10^6$ cycles), applied load (50 N), sliding amplitude (75 μm), sliding frequency (10 Hz) and oxygen content (10^{-6} wt%). The temperature was varied in the range between 450 and 550°C.

As shown in Fig. 6.40 and 6.41, for all the materials, volume loss and fretting depth increase with the increasing temperature.

GESA-T91 is characterized by the lowest fretting wear and the smallest increase with the temperature. From 450 to 550°C, volume loss and fretting depth increase from 0.0064 to 0.018 mm³ and from 12 to 17 µm (from 2.4 to 3.4 % penetration) respectively. Besides, wear rates and specific wear coefficients measured for GESA-T91 are up to an order of magnitude lower that the one of T91 and 15-15Ti (Fig. 6.42 and 6.43).

Friction pairs of 15-15Ti showed the biggest fretting damage up to 500°C. The temperature increase from 450 to 550°C causes an increase of volume loss and fretting depth by a factor ~ 2.5 (up to 0.113 mm³) and ~ 1.5 (up to 85 µm= 17 % penetration) respectively. T91 is the material most affected by the temperature increase and at 550°C the fretting damage of this alloy exceeds the one measured on 15-15Ti.

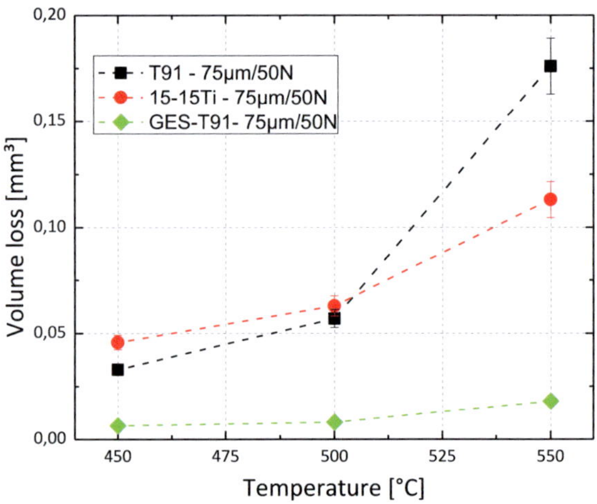

Fig.6.40. Graph of volume loss versus temperature for friction pairs of T91, 15-15Ti and GESA-T91. Test parameters: 75 µm, 50 N, 10 Hz, 150 h, $5.4 \cdot 10^6$ cycles, 10^{-6} [O] wt%.

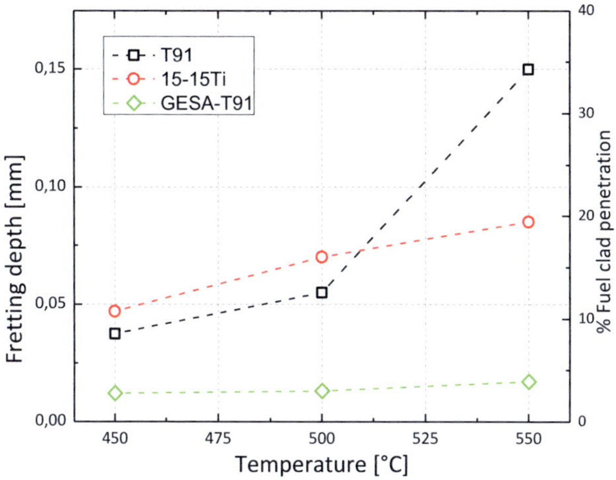

Fig.6.41. Graph of fretting depth versus temperature for friction pairs of T91, 15-15Ti and GESA-T91. Test parameters: 75 μm, 50 N, 10 Hz, 150 h, $5.4\cdot10^{6}$ cycles, 10^{-6} [O] wt%.

For T91, from 450 to 550°C, the volume loss increases by a factor 5 (up to 0.176 mm³) and the fretting depth increases by a factor 4 (up to 150 μm = 30 % penetration).

At 450 and 500°C, friction pairs of 15-15Ti have the highest values of wear rate and specific wear coefficient. However they are close to the ones of T91.

At 550°C wear rate and specific wear coefficient of T91 increase rapidly and reach higher values than the ones of 15-15Ti. At this temperature wear rate (Fig. 6.42) and specific wear coefficient (Fig. 6.43) for T91 and 15-15Ti are $\sim3.2\cdot10^{-7}$ mm^3s^{-1}/$\sim4.3\cdot10^{-15}$ $m^3N^{-1}m^{-1}$ and $\sim2.3\cdot10^{-7}$ mm^3s^{-1} /$\sim2.3\cdot10^{-15}$ $m^3N^{-1}m^{-1}$ respectively.

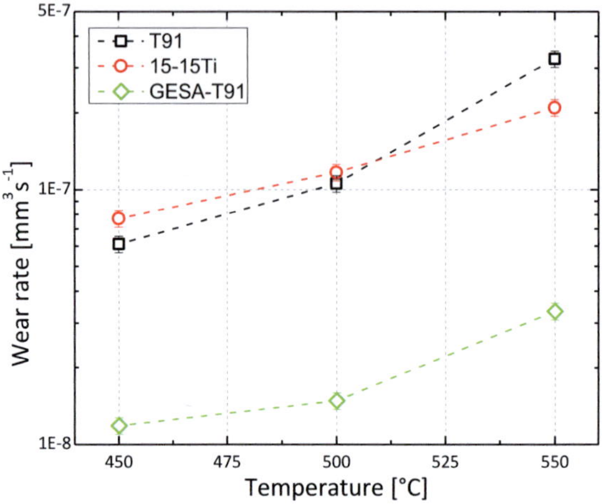

Fig.6.42. Graph of wear rate versus temperature for friction pairs of T91, 15-15Ti and GESA-T91. Test parameters: 75 μm, 50 N, 10 Hz, 150 h, $5.4 \cdot 10^6$ cycles, 10^{-6} [O] wt%.

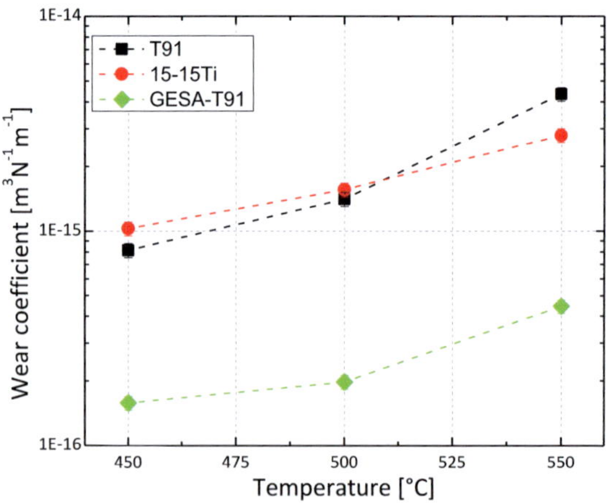

Fig.6.43. Graph of the specific wear coefficient versus temperature for friction pairs of T91, 15-15Ti and GESA-T91. Test parameters: 75 μm, 50 N, 10 Hz, 150 h, $5.4 \cdot 10^6$ cycles, 10^{-6} [O] wt%.

Metallographic observations

The temperature influences also the corrosion behaviour of the friction pairs, especially of T91 and 15-15Ti. The oxidation process, enhanced by the temperature increase, generally favours the growth of thicker oxide scales which, in similar manner for all the tested materials, are removed by the fretting action.

For the friction pairs of T91 after 150 h the temperature increase does not cause remarkable changes or dissolution attacks in the fretted area. However the temperature increase has a remarkable effect on the oxidation process. Indeed, in the areas not affected by fretting the protective oxide scale thickness increases with the temperature. The duplex oxide scale (outer layer of Fe_3O_4 + inner layer of Fe-Cr spinel oxides) forming on the steel surface after exposure in Pb with 10^{-6} wt% of oxygen is about 4, 7 and 10 μm at 450, 500 and 550°C respectively. Additionally, after fretting test in Pb at 550°, beneath the Fe-Cr spinel oxides scale, an internal oxidation zone (IOZ) of about 3 μm was also noticed. Such observations are supported by the SEM/EDX line scan of Fig. 6.44. The EDX curves of O (light blue), Fe (red) and Cr (pink) suggest the presence of a duplex oxide scale and of an internal oxidation zone. According to the EDX line analysis, the light spots in the SEM micrograph in BSE mode of Fig. 6.44 consists of Pb that penetrated the magnetite layer (Fe_3O_4).

The friction pairs of 15-15Ti are affected by the temperature increase mainly concerning the dissolution process in the fretted area. As shown in Fig. 6.45, after fretting tests at 150 h in Pb with 10^{-6} wt% of oxygen for all the temperatures tested (400, 500 and 550°C) a similar compacted scale of about 15 μm forms. Such scale, as explained in paragraph 6.1.1, consists of oxidized and metal debris, Ni depleted and mixed with Pb. However, at 500 and 550°C Pb penetration and dissolution attack, with consequent Ni depletion (as illustrated in the SEM/EDX image and line scan of Fig. 6.8),

affect also the bulk material. The dissolution depth at 550°C (~ 35 μm) is more significant than at 500°C (~ 20 μm).
In the not-fretted area, the steel surface is generally covered by a protective oxide scale (Fe-Cr spinel) and no dissolution attack was observed.
On the friction pairs of GESA-T91 no particular effect related to the temperature increase was noticed.

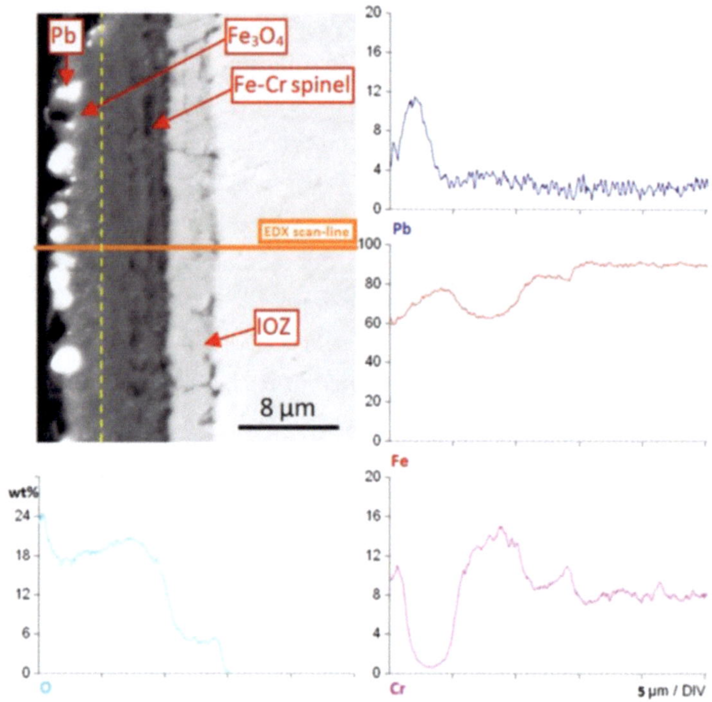

Fig.6.44. SEM image in BSE mode and EDX line scan of a T91 specimen cross-section after fretting test in Pb with 10^{-6} wt% of oxygen at 550°C for 150 h. The EDX lines suggest the presence of a duplex oxide scale and of IOZ beneath. The light spots are Pb penetrating the oxide scale. Test parameters: 50 N, 75 μm, 10 Hz.

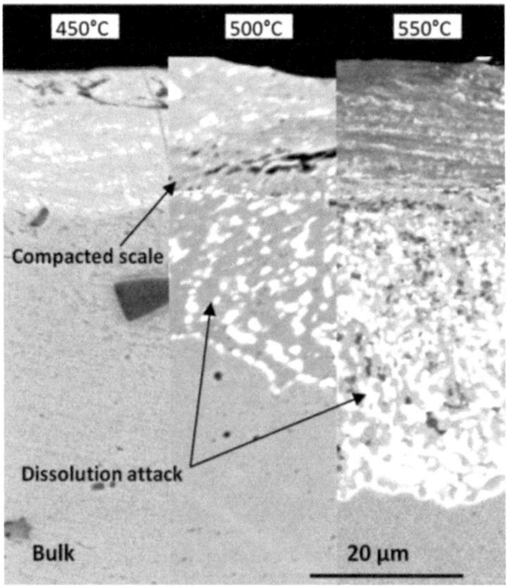

Fig.6.45. SEM images in BSE mode of 15-15Ti specimens after fretting tests in Pb at 450, 500 and 550°C for 150 h. Dissolution depth increase with temperature. Test parameters: 50 N, 75 µm, 10 Hz.

6.1.6 *Influence of a pre-formed oxide scale*

The influence that a pre-formed oxide scale might have on the fretting process was evaluated executing a fretting test on friction pairs pre-exposed in liquid Pb with oxygen saturation for 200 h at 500°C.The pre-exposure conditions (time, temperature and oxygen content) were expected to be adequate to develop on the alloys surfaces the double layered oxide scales of few µm for T91 and 15-15Ti and the nanometric alumina scale on GESA-T91. By that, when the fretting test starts, on the contact surfaces of the friction pairs, a pre-formed oxide scale exists at the starting of the fretting action. The fretting test with pre-exposure was performed at following conditions: 50 N, 75 μm, 10 Hz, 450°C, 10^{-6} wt% of oxygen in liquid Pb, 150 h, and $5.4 \cdot 10^{6}$ cycles.

According to the diagrams of Fig. 6.46 and 6.47, the pre-exposure causes reductions in volume loss and fretting depth for all the tested material. However, the most evident reduction in volume loss and fretting depth observed for the friction pair of 15-15Ti. The volume loss decreases by about 7.8 % for GESA-T91, 18.2 % for T91 and 43 % for 15-15Ti. Similar observations can be made for the fretting depth. The pre-exposure causes the fretting depth to decrease from 12 to 11 μm (-8.4 %) for GESA-T91, from 37.5 to 24 μm (-36 %) for T91 and from 47 to 22 μm (-64 %) for 15-15Ti.

The reduction in volume loss and fretting wear for the friction pair pre-exposed results in lower wear rates and specific wear coefficients (Fig. 6.48 and 6.49). Wear rate and specific wear coefficient reduction due to the pre-exposure for T91 and GESA-T91 is less evident than for 15-15Ti.

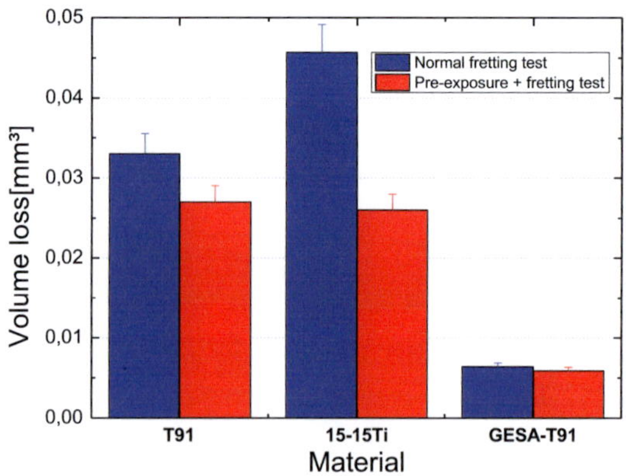

Fig.6.46. Graph of volume loss for friction pairs of T91, 15-15Ti and GESA-T91 submitted to fretting test after pre-exposure and without pre-exposure. Test parameters: 75 μm, 50 N, 10 Hz, 450°C, 10^{-6} [O] wt%, 150 h, $5.4 \cdot 10^{6}$ cycles.

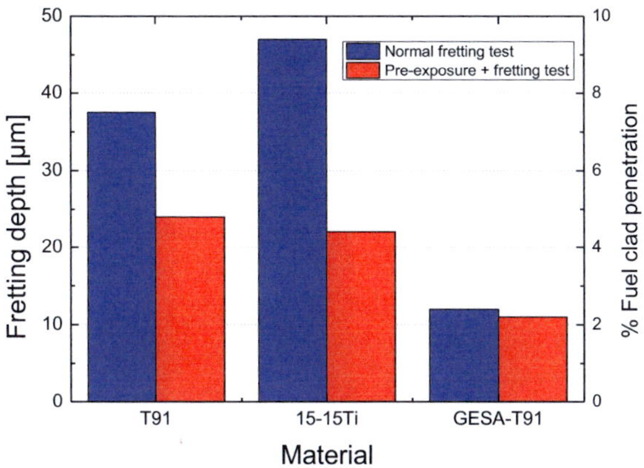

Fig.6.47. Graph of fretting depth for friction pairs of T91, 15-15Ti and GESA-T91 submitted to fretting test after pre-exposure and without pre-exposure. Test parameters: 75 µm, 50 N, 10 Hz, 450°C, 10^{-6} [O] wt%, 150 h, $5.4 \cdot 10^{6}$ cycles.

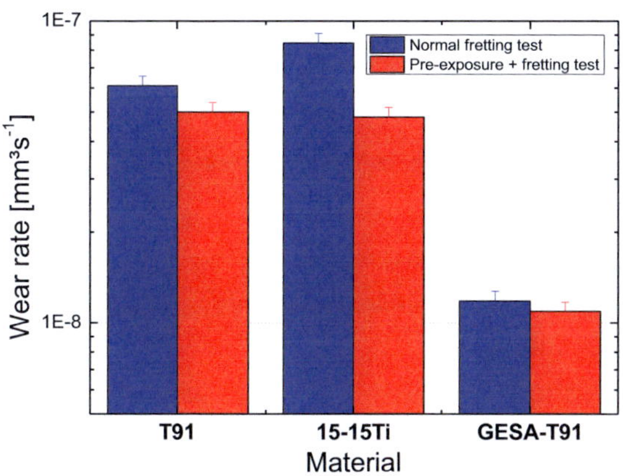

Fig.6.48. Graph of wear rate for friction pairs of T91, 15-15Ti and GESA-T91 submitted to fretting test after pre-exposure and without pre-exposure. Test parameters: 75 µm, 50 N, 10 Hz, 450°C, 10^{-6} [O] wt%, 150 h, $5.4 \cdot 10^{6}$ cycles.

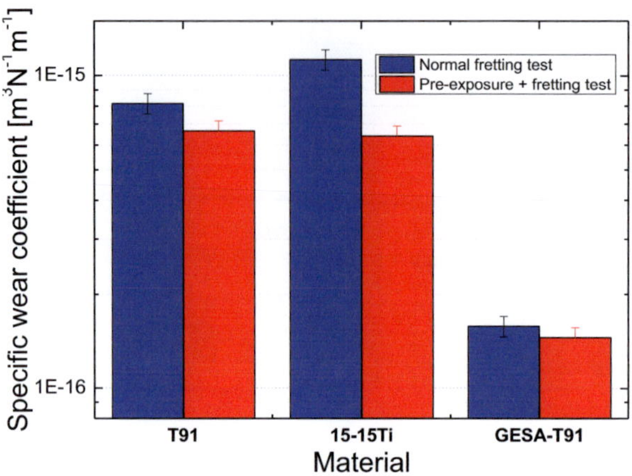

Fig.6.49. Graph of specific wear coefficient for friction pairs of T91, 15-15Ti and GESA-T91 submitted to fretting test after pre-exposure and without pre-exposure. Test parameters: 75 μm, 50 N, 10 Hz, 450°C, 10^{-6} [O] wt%, 150 h, $5.4 \cdot 10^6$ cycles.

Metallographic observations

The pre-exposure at 500°C for 200 h favors the development of oxide scales as previously described in the paragraph 6.1.5, where the role of the temperature is discussed.

The presence of an oxide scale formed before starting the fretting test has an influence also on the formation of the interposing layer of compacted debris. Indeed, after 150 h test ($5.4.10^6$ cycles) on pre-exposed friction pairs, the interposing layer consists mainly in oxidized debris and the layer of compacted metal debris mixed with Pb was not observed, which is confirmed by the SEM micrograph and the EDX line scan analysis of Fig. 6.50. This image, which refers to the friction pair of 15-15Ti, shows the formation of a 5-10 μm thick compacted scale made of oxidized debris mixed with Pb.

This observation is supported by the EDX analysis lines of O, Cr, Fe and Pb. Besides, almost no Ni is detected in the compacted scale but Ni enrichment can be noticed at the interface compacted scale / bulk. Underneath the compacted scale, in the bulk, the EDX analysis of the main alloying elements reveals values near to the nominal composition of the 15-15Ti steel.

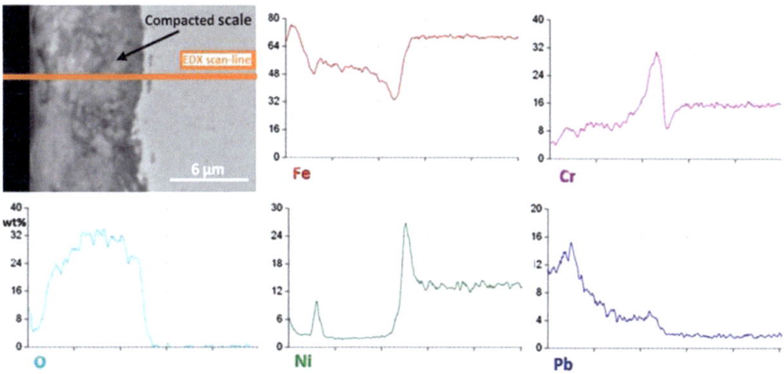

Fig.6.50. SEM image in BSE mode and EDX line scan of the compacted scale formed on 15-15Ti after pre-exposure and fretting test. The EDX lines indicate a compacted scale made of oxidized debris and Ni enrichment at the compacted scale/bulk interface. Test parameters: 75 µm, 50 N, 10 Hz, 450°C, 10^{-6} [O] wt%. Pre-exposure in oxygen saturated Pb for 200 h at 500°C

6.1.7 Influence of Ni dissolved in molten lead

The SEM/EDX analysis of friction pairs of 15-15Ti submitted to fretting test, highlighted that in most of the testing conditions the compacted scale forming on the fretted area is affected by dissolution that concerns especially Ni. In particular experimental conditions (see paragraph 6.1.2 and 6.1.5), starting from the compacted scale, the dissolution spreads inwards and affects the bulk material. Dissolution was not noticed for the other tested materials (T91 and GESA-T91). Thus, to mitigate the dissolution and to evaluate the eventual effect that this might have on the fretting wear/corrosion process, a set of fretting test in Pb with dissolved Ni (Ni-enriched Pb) was performed on friction pairs of 15-15Ti. Fretting tests were carried out in the following conditions: 50 N, 75 μm, 10 Hz, 450°C, 10^{-6} wt% of oxygen in Pb and for exposure times of 150, 312 and 930 h. The results of these tests are plotted together with data of fretting tests performed in pure Pb on friction pairs of T91 and 15-15Ti.

Fig. 6.51 highlights a reduction in the volume loss, for the friction pairs of 15-15Ti tested in Ni-enriched Pb, of about 50 % for all the exposure times. In this testing conditions, for 15-15Ti also the fretting depth (Fig. 6.52) is decreased by 54 % (from 47 to 22 μm) after 150 h and by 41 % (from 110 to 66 μm) after 930 h.

Besides, as depicted in the diagrams of Fig. 6.51 and 6.52, friction pairs of 15-15Ti tested in Ni-enriched Pb are even less susceptible to fretting wear than T91. For example, fretting depths plotted for 15-15Ti in Fig. 6.53 are from 10 to 40 % lower than the ones measured on friction pairs of T91.

The reduction of volume loss measured after tests performed in Ni-enriched Pb correspond to reduced wear rates and specific wear coefficients (diagrams of Fig. 6.53 and 6.54).

After 150 h test the fretting wear and specific wear coefficient from pure lead to Ni-enriched Pb decreases by about 50 %.

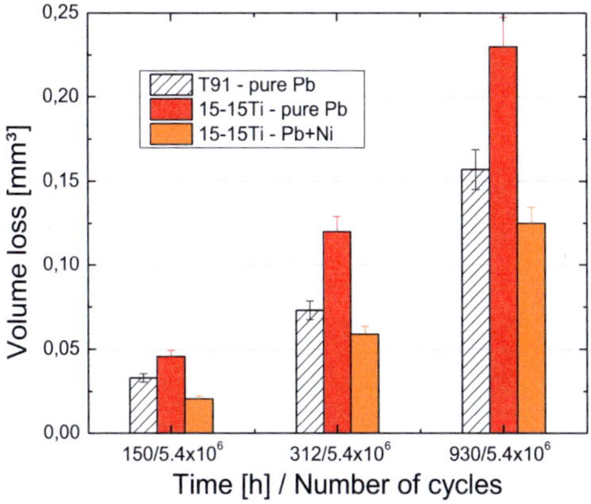

Fig.6.51. Diagram of volume loss for friction pairs of 15-15Ti submitted to fretting test in Ni enriched lead and in pure lead. The values of T91 are also reported for comparison. Test parameters: 75 µm, 50 N, 10 Hz, 450°C, 10^{-6} [O] wt%.

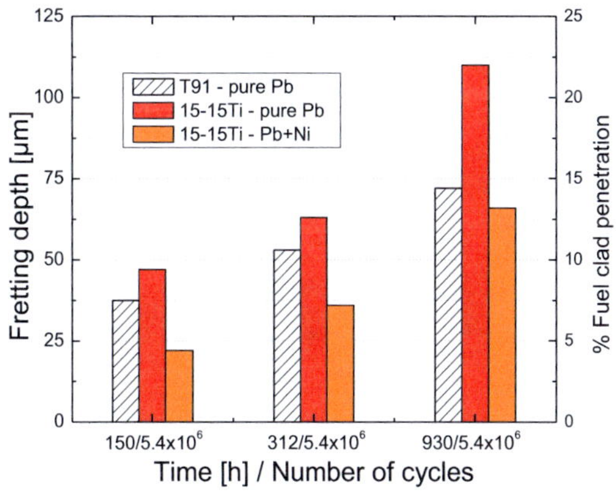

Fig.6.52. Graph of fretting depth for friction pairs of 15-15Ti submitted to fretting test in Ni enriched lead and in pure lead. The values of T91 are also reported for comparison. Test parameters: 75 µm, 50 N, 10 Hz, 450°C, 10^{-6} [O] wt%.

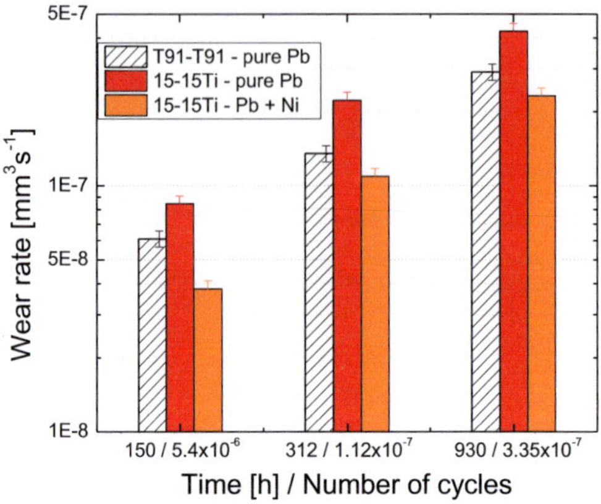

Fig.6.53. Graph of wear rate for friction pairs of 15-15Ti submitted to fretting test in Ni enriched lead and in pure lead. The values of T91 are also reported for comparison. Test parameters: 75 μm, 50 N, 10 Hz, 450°C, 10^{-6} [O] wt%.

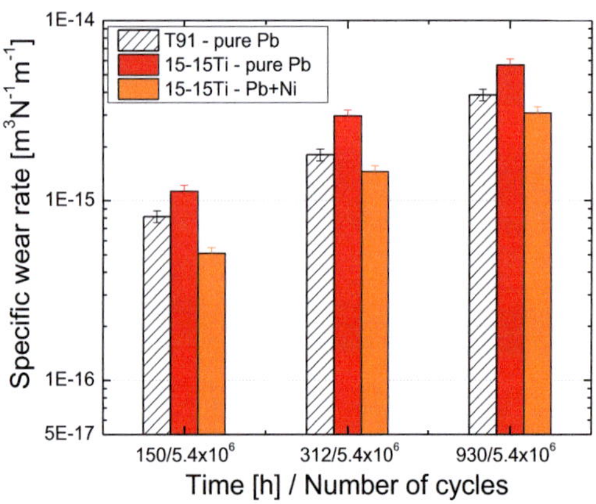

Fig.6.54. Graph of specific wear coefficient for friction pairs of 15-15Ti submitted to fretting test in Ni enriched lead and in pure lead. The values of T91 are also reported for comparison. Test parameters: 75 μm, 50 N, 10 Hz, 450°C, 10^{-6} [O] wt%.

Metallographic observations

The presence of Ni dissolved in liquid Pb has also a remarkable influence on the corrosion behavior of the friction pairs of 15-15Ti. As shown in Fig. 6.55, the dissolution attack affecting the bulk material after the fretting test of 930 h at 450°C and 10^{-6} wt% of oxygen in solution is deeper and more extended in the case of the test performed in pure Pb than for the test performed in Ni-enriched Pb. In pure Pb dissolution reaches 20-30 µm and is extended on most of the fretted surface (left side Fig. 6.55). Conversely, in the case of the tests performed in Ni-enriched Pb (right side Fig. 6.55), the dissolution attack is visible only in few localized places and the maximum dissolution depth is around 15-20 µm. The compacted scale seems substantially the same in the two cases, the only noticeable difference is the more significant thickness of the compacted scale (up to 30-35 µm) forming during the test performed in Ni enriched Pb.

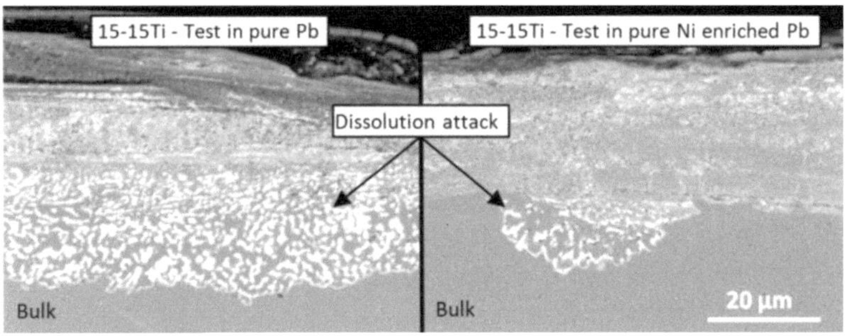

Fig.6.55. SEM image in BSE mode of the dissolution attack and the compacted scale characterizing friction pairs of 15-15Ti after fretting test of 930 h at 450°C and 10^{-6} wt% in pure Pb (left side) and Ni-enriched Pb (right side).Test parameters: 75 µm, 50 N, 10 Hz.

6.2 Discussion of the results

6.2.1 General effects of liquid lead

The outcomes of the experimental campaign carried out in molten lead highlighted that fretting in liquid Pb is less severe than in normal atmosphere. Fig. 6.56 displays that with the same test parameters, the volume loss resulting from tests performed on friction pairs of T91 in air at room temperature (blue squares), is about from 2.5 to 5 times bigger than in liquid Pb at 450°C (red circles). Even at 300 (pink diamond) and 450°C (light blue triangles) tests in air resulted in fretting wear up to 2-3 times bigger than in liquid Pb at 450°C.

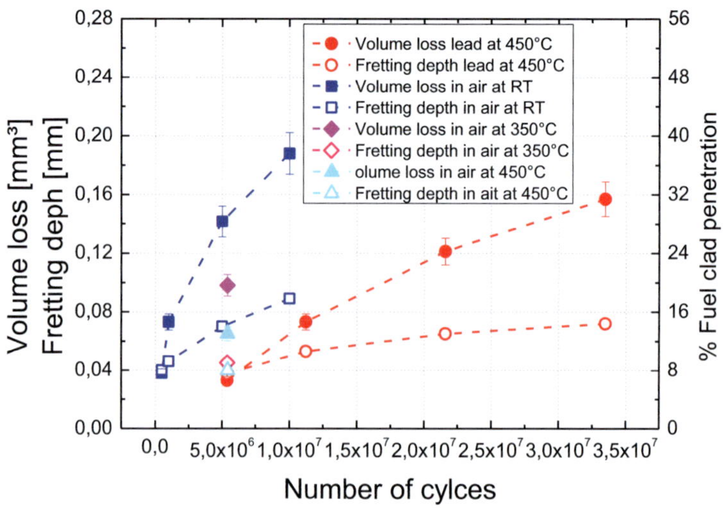

Fig.6.56. Graph of volume loss and fretting depth versus number of cycles for friction pairs of T91 tested in air at room temperature, at 300 and 450°C and in Pb at 450°C. Test parameters: 50 N, 75 μm, 10 Hz (Pb) / 20 Hz (air).

Similar observation can be made for penetration depth (Fig. 6.56), wear rate (Fig. 6.57) and specific wear coefficient (Fig. 6.57).

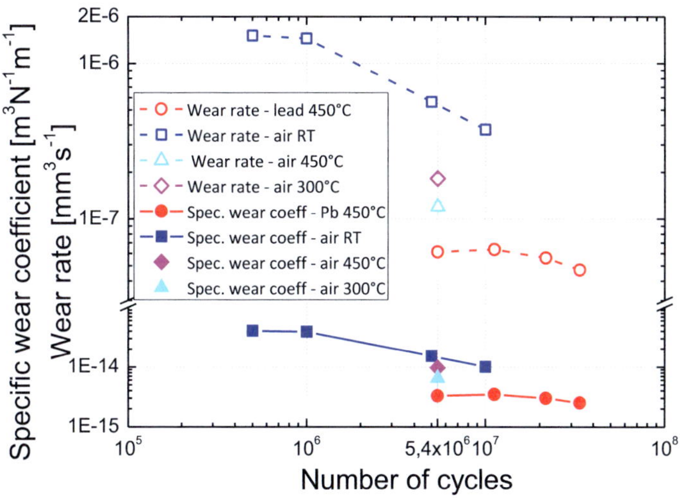

Fig.6.57. Graph of wear rate and specific wear coefficient versus number of cycles for friction pairs of T91 tested in air at room temperature, at 300°C, at 450°C and in Pb at 450°C. Test parameters: 75 μm, 50 N 20 Hz (air) / 10 Hz (Pb).

Such results can be explained considering that in the contact area liquid Pb acts as a lubricant. At 450°C the dynamic viscosity (η) of molten Pb is $1.96 \cdot 10^{-3}$ Pa·s, which corresponds to a kinematic viscosity υ (υ = η/ρ with ρ = density) of $1.81 \cdot 10^{-7}$ m^2s^{-1} (m^2s^{-1} = Stokes) [2]. Such values are slightly lower compared to the ones of commercial oils for car engines. Thus, molten Pb, forming a liquid film between the sliding loaded surfaces of specimen and counter-specimen, lowers the coefficient of friction.

In Fig. 6.58, values of the tangential force experimentally measured for friction pairs of T91 tested in Pb and in air are plotted for different applied loads. The values of tangential force, detected for the particular FRETHME setup, suggest a friction coefficient of around 0.53 and 0.28 for fretting contact in air and in molten Pb respectively.

The presence of Pb influences also the composition of the compacted scale and the corrosion processes that affect the friction pairs.

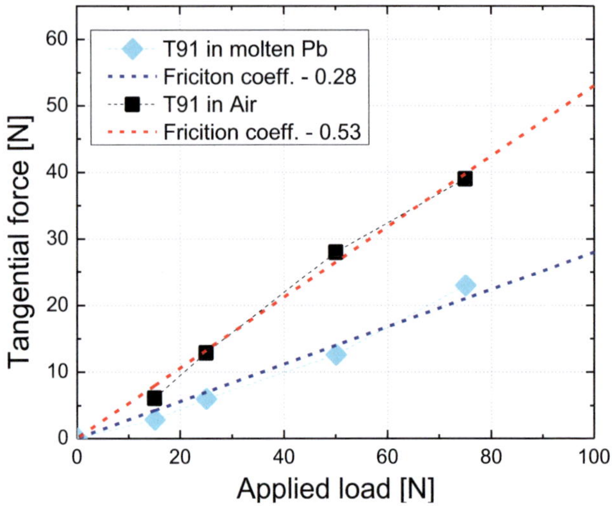

Fig.6.58. Graph of measured tangential force versus applied load for friction pairs of T91 tested in air at room temperature and in Pb at 450°C. Test parameters: 50 N, 75 μm, 10 Hz (Pb) / 20 Hz (air).

6.2.2 Lead effects in non-fretted areas

Fretting test in liquid Pb were performed in reactor relevant conditions in terms of temperature (450-550°C) and oxygen content (~10^{-6} wt%). In this environment and for the exposure times adopted, oxide scales (protective against dissolution) are expected to grow on the surface of specimens and counter-specimens of the materials under investigation [2, 41, 46, 51] (see paragraph 1.5.). Accordingly, on the surfaces of the friction pairs exposed to molten Pb but not affected by fretting, oxide scales were detected and analysed. Thickness and composition of such scales are coherent with the data reported in the literature [2, 39-55] and addressed in Chapter 1. This agreement indicates that fretting tests were accomplished appropriately in terms of temperature and oxygen content.

Double layered oxide scales (duplex) were observed on friction pairs of T91. The EDX analysis confirmed that such scale consists of an outer layer of Fe oxides (most probably magnetite - Fe_3O_4) and an inner layer of Fe-Cr oxides (most probably spinel oxides). The oxide scale thickness varies with temperature and exposure time. At 450°C with 10^{-6} wt% of oxygen liquid Pb, the thickness of the duplex oxide scale reaches 12 μm after 930 h, following a parabolic growth in time (Fig.6.59) that agrees with the literature records [2, 39-55].

At 500 and 550°C the oxide scale is 7 and 10 μm thick already after 150 h. Additionally, after 150 h exposure in molten Pb at 550°C, the formation of an internal oxidation zone was noticed. The higher temperature favours the diffusion process. As described in paragraph 1.5.3 and schematically shown in Fig. 1.9, the faster outwards diffusion of Fe cations, creating more vacancies in the steel matrix, and the inwards oxygen diffusion create the conditions for internal oxidation.

The friction pairs of 15-15Ti, are characterized in the not fretted areas by Fe-Cr oxides scale which, according to the information provided in paragraph 1.5, is most probably a spinel type oxide. Temperature and exposure time influence the oxide scale thickness that is generally lower than the duplex oxide scale forming on T91. As a result, at 450°C with 10^{-6} wt% of oxygen in liquid Pb the oxide scale thickness after 930 h exposure is about 4 μm (Fig. 6.59). Also for 15-15Ti, in agreement with the literature [2, 39-55], the oxide scale growth follows a parabolic law.

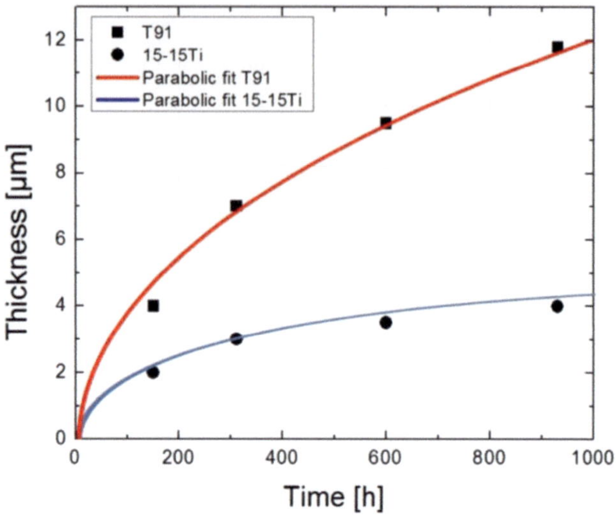

Fig.6.59. Plot of the oxide scale thickness versus the exposure time for T91 and 15-15Ti exposed to liquid Pb with 10^{-6} wt% of oxygen at 450°C.

The oxide scale growth on the surfaces of 15-15Ti exposed to liquid Pb is slower than for T91 due to the higher Cr content of the 15-15Ti (~15 vs ~9 wt%) and its austenitic microstructure. As reported in [44, 49], an adequate Cr content can promote the formation of a Cr-rich transient oxide scale on the surface of steels exposed to oxygen containing media. Through the Cr-rich oxide scale, the diffusion rates of Fe cations (outwards) and oxygen (inwards) are lowered so that internal (Fe-Cr oxides and IOZ) and external (Fe-oxides) oxidation rates are reduced. Besides, due to the coarser grain size, the 15-15Ti austenitic structure is characterized a priori by lower diffusion rate than the T91 martensitic structure. The grain size is of great importance with regard to the species diffusion in particular at relatively low temperatures when the diffusion rate is already low [183]. As a matter of fact, grain boundaries are fast diffusion paths along which the alloying elements can move in the metal matrix. Thus, the finer the grain size the higher is the diffusion coefficient [184, 185]. As a result, thinner oxide scales are expected for 15-15Ti (Fig. 6.59) due to its larger grains size.

In the not-fretted areas, the GESA –T91 specimens are covered by an alumina scale. Such scale, compared to Fe and Fe-Cr oxides scale, is characterized by a significantly lower growing rate [53, 186, 187]. This thin oxide scale could not be directly detected with the characterization methods applied for this work. Indeed, the relative short exposure times in liquid Pb at the tested temperatures are most likely only sufficient to develop a nanometric alumina scale. However, the presence of the alumina scale (Al_2O_3) for the tested experimental conditions was proved in previous works [53, 186, 187] and it is expected for thermodynamic and composition related reasons. As shown by the Ellingham diagram reported in Fig. 1.7 (paragraph 1.5.2), for temperatures and oxygen content tested in this work (450-550°C / 10^{-6} wt% [O]), aluminium oxide (Al_2O_3) is the most favourable oxide from the thermodynamic point of view compared to Fe, Cr and Pb oxides. Besides, the LPPS deposition of FeCrAlY powders on the T91 substrate and the following electron beam treatment (GESA) lead to the formation of a 20-30 µm thick surface alloyed layer that contains around 7 wt% of aluminium. As confirmed by previous studies [53, 186, 187], this Al content is sufficient to develop a thin protective alumina scale on the GESA treated T91 surface exposed to liquid Pb.

The oxide scales forming on the three alloys under investigation can act as a solid lubricant in the fretted area [71, 72, 89, 92, 106-114]. Such characteristic was highlighted by fretting tests performed on friction pairs covered by a pre-formed oxide scale (due to pre-exposure in Pb). For all the materials tested the presence of an oxide scale reduced the fretting wear in some cases up to 50-60 % (Figs. 6.46 - 6.49). Pre-exposed 15-15Ti showed volume loss and fretting depth similar to the ones of T91. Thus, 15-15Ti is the material that gains most from the presence of an oxide scale in the fretted area. This aspect underlines that the presence of an oxide scale can delay the wear process.

6.2.3 Nature of the fretted area

As detailed in chapter 5, in the area of contact between specimen and counter-specimen, where the fretting process takes place, protective oxide scales (Fe, Fe-Cr oxide scale) and corrosion barrier (GESA alloyed layer) are removed, fretting debris are generated and an interposing scale of compacted debris forms. The microstructural analysis (e.g. Figs. 6.3, 6.7 and 6.11) emphasized the absence of any specific ordered structure in the interposing scale. This confirms that the interposing scale is made of debris generated during the fretting action that are confined and compacted in the contact area due to the short sliding amplitude (< 200 μm) and to the contact pressure (between ~5 and ~40 MPa) [56, 58, 98]. Additionally, the SEM/EDX analysis of the compacted scale highlighted the presence of two layers, an outer one, consisting in a breakage of Pb and oxides, and an inner one made up mostly of Pb and metal debris. The composition of the compacted interposing scale varies with kind of steel, temperature and exposure time. Fe and Fe-Cr oxides were detected in the outer layer of the interposing scale forming on T91 and 15-15Ti; whereas a mixture of Fe, Cr and Al oxides was noticed for friction pairs of GESA-T91. For all the steels, the oxidation and the amount of Pb in the compacted scale increase with duration and temperature of the fretting test. Higher temperature causes faster oxidation; longer test duration gives more time to liquid Pb to penetrate the compacted scale [48, 52, 181, 188, 189]. For T91 and 15-15Ti, the thickness of the compacted scale is in most of the cases up to 20 μm, while for GESA-T91 it is up to 15 μm. The compacted scale might break away during the fretting process or during the cross-section cutting procedure; as a consequence it was not possible to determine a clear relation between its thickness and the fretting parameters. However, it was generally observed that short sliding amplitudes (≤ 35 μm) lead to the formation of thinner compacted scales mainly consisting of oxidized debris. A short amplitude favors slow oxide scale removal (e.g. Fig. 6.38) and debris retention that delay the wear of the bulk material.

In the specific case of the 15-15Ti steel, the fretted areas are characterized by dissolution attack and consequent substantial Ni depletion. The dissolution attack, depending on exposure time and temperature, affects only the compacted scale or spreads into the bulk material. However it generally does not affect the not fretted area where protective oxide scales develop.

For all the testing conditions, the compacted scale is almost completely Ni depleted even in case that dissolution attack does not affect the bulk material. Molten Pb penetrates the compacted scale and it mixes with metal debris. In this situation, dissolution of Ni is favored. In the temperature range 450-550°C, Ni solubility in Pb is up to 0.6 wt% (Fig. 1.6) [2]. Additionally, due to the high surface area to volume ratio (S/V), dissolution of particles abraded by the sliding movement is favored compared to dissolution of bulk material [190, 191].

Molten Pb, spreading through the scale of compacted debris, reaches the underlying bare 15-15Ti surface. Here, the absence of a protective oxide scale (removed by fretting) and a sufficient exposure time in liquid Pb create favorable conditions for the inwards spreading of dissolution. As a result, after fretting tests longer than 312 h, dissolution affects also the bulk material with depth and extent that increase with time (~20 μm deep after 930 h at 450°C) and temperature (~35 μm deep after 150 h at 550°C). After the fretting test of 600 h, cracks most probably generated by the cyclic loading (delamination theory / fatigue cracking, see Chapter 2) were detected [58]. These cracks, besides affecting the mechanical properties, act as preferential channel for liquid Pb penetration, increasing the spreading rate of dissolution through the bulk material.

No remarkable dissolution effects were noticed in the compacted scale of T91. Fe and Cr, the main T91 steel alloying elements, have much lower solubility (4 -5 orders of magnitude) in Pb than Ni.

Partial Al dissolution was detected in the compacted scale forming on the GESA-T91. In the temperature of the fretting test Al solubility in Pb is up to 0.06 wt%. Another peculiarity of GESA-T91 is that the fretting action gradually consumes the Al containing surface alloyed scale, so that after 930 h and 3.35 10^7 fretting cycles most of the alloyed layer is removed (Fig. 6.21). For shorter tests (i.e. < 600 h / $2.16 \cdot 10^7$ cycles) the surface alloyed layer is thinned but still entirely covering the T91 substrate. Thus, in case of interruption of the fretting motion, the thin protective alumina scale can still grow even after $1\text{-}2 \cdot 10^7$ cycles.

6.2.4 Fretting wear evolution

For all the tested materials, given a fixed set of fretting parameters, fretting wear increases with the time/number of cycles. In Fig. 6.60, the volume loss is reported as function of the time and, considering the measurements error (7.5 % - see paragraph 4.4.1), suitable fittings are also plotted. A parabolic rate seems to be adequate for friction pairs of T91 and 15-15Ti (in Fig. 6.60, black line and red line respectively). For these materials, the first ~ 300 h/~10^7 cycles are characterized by faster volume removal (higher wear rate) than for longer exposure / higher number of cycles. As shown in Fig. 6.14, for T91 and 15-15Ti, the wear rate (in mm^3s^{-1}) is higher in the first 10^7 cycles than afterwards.

The parabolic growth of fretting wear with time seems to be is a peculiarity of fretting of steel. Indeed, similar trends, characterized by an initial higher removal rate, are shown in this work for fretting of T91 in normal atmosphere and in previous research activities focused on fretting of austenitic steels in gaseous and aqueous media [121].

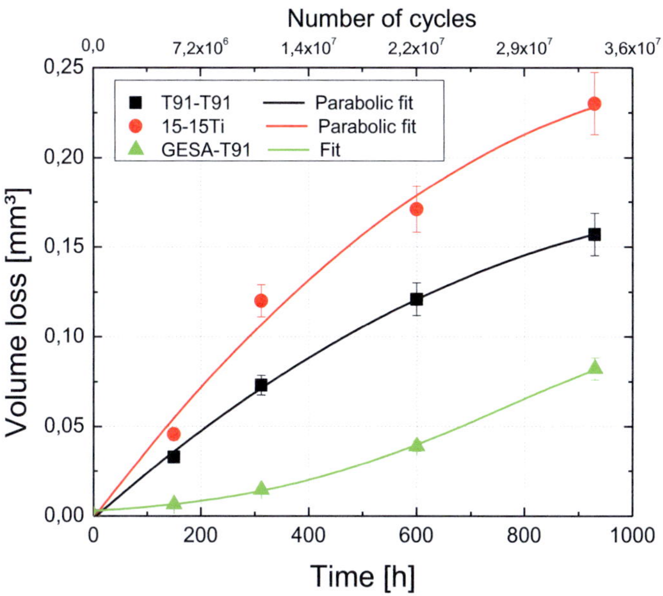

Fig.6.60. Graph of volume loss versus number of cycles for friction pairs of T91, 15-15Ti and GESA-T91. Test parameters: 50 N, 75 µm, 10 Hz, 450°C, 10^{-6} wt% [O].

The higher fretting susceptibility of 15-15Ti compared to T91 is most likely related to dissolution/oxidation processes occurring in liquid Pb that interact with fretting. On one hand, Pb penetration and Ni depletion affecting the bulk of 15-15Ti (especially for long term fretting test) can facilitate the material removal (dissolution enhanced fretting). This idea is supported by the fact that tests performed with Ni-enriched Pb, in which the Ni dissolution rate is lower, resulted in a reduced fretting wear (even lower than for T91) compared to tests performed in pure Pb (Figs. 6.51 – 6.54). In this condition, the SEM/EDX analyses showed fretted areas less severely affected by dissolution (Fig. 6.55).

On the other hand, as aforementioned, oxide scales forming on the steel surface act as a kind of solid lubricants that can mitigate fretting. In partic-

ular, the magnetite (Fe_3O_4), which constitutes the outer layer of the duplex oxide scale forming on the surface of T91 exposed to liquid Pb, was found to have the highest mitigating power among the other Fe oxides [130, 131]. Additionally, T91 steel is characterized by a higher oxidation rate than 15-15Ti steel. The duplex oxide scale growing on T91 surface is thicker and more effective as solid lubricant than the Fe-Cr spinel forming on 15-15Ti steel. Moreover, the magnetite layer (Fe_3O_4) develops on the original steel-liquid metal interface while the Fe-Cr spinel layer forms in the bulk due to the inwards diffusion of oxygen (Fig. 1.9 in paragraph 1.5.3). As a consequence, the initial removal of magnetite does not reduce the original components cross-section and the fretting wear of the bulk material is delayed.
The GESA-T91 (green triangles in Fig. 6.60) showed a different time dependence compared to T91 and 15-15Ti. For this material, the wear rate is lower than for original T91 but increases remarkably for fretting tests longer than 312 h ($1.1 \cdot 10^7$ cycles). For longer times it tends to approach wear rates characteristic of T91. The lower wear rate measured during the first 150 h can be associated to the thin slowly growing alumina scale and to lower surface finishing characterizing the GESA-T91 specimens. On one hand the alumina scale is mechanically and chemically more stable than the Fe and Fe-Cr oxides forming on T91 and 15-15Ti [52-54, 185]. On the other hand, the higher roughness (Rq = ~5 μm) than e.g. original T91 (Rq = < 2 μm), leads to a lower number of contacts points between specimen and counter-specimen and also reduces (at least in the first stage of the fretting process) the surface affected by the fretting ploughing action (see paragraph 2.3.7). The effects of oxide scale and roughness likely run out after the first 150 h so that an increase in wear rate can be expected. However, the hardness 60-80 % higher compared to the unmodified T91 and smaller grain size [53, 186] typical of the GESA alloyed layer assure a higher wear resistance than original T91. The fretting process removes gradually the GESA alloyed layer and, when the bare T91 substrate is

exposed to the liquid Pb and to the fretting action (~$2 \cdot 10^7$ fretting cycles - see Figs. 6.9, 6.10 and 6.11), the fretting behaviour of GESA-T91 becomes the same as unmodified T91 (in Fig. 6.60, the curves of these two materials reach the same slope).

6.2.5 Combined effects of load and amplitude – The fretting regimes

For friction pairs of T91, the fretting wear (volume loss and fretting depth) increases with the applied load up to a certain point and then it decreases (Figs. 6.22 – 6.25). Loads in the range 15-75 N were applied with different fixed imposed amplitudes. The results showed that the value of applied load corresponding to the maximum fretting wear (turning point in plot fretting wear-load) changes in dependence of the imposed sliding amplitude. Maximum fretting wear was detected with applied loads of 25, 50 and higher than 75 N for imposed amplitudes of 35, 75 and 165 μm respectively.

The discussion on the role of applied load and sliding amplitude for fretting in normal atmosphere pointed out that these two variables affect the wear process mainly by influencing debris retention and by determining the fretting regime. Also for fretting tests performed in molten Pb, the influence of load and sliding amplitude is here discussed in these terms.

The presence of solidified Pb on the fretted surface complicates the surface analysis. However, it is reasonable to assume that, similarly to fretting in normal atmosphere, high loads (high local pressure) and short amplitudes favour the debris retention in the contact area. On the contrary, long amplitude and low loads favour the debris release from the contact area.

Debris retention favours the formation of a compacted interposing scale that withstands the contact load, tends to separate specimen and counter-specimen surfaces and acts as a kind of solid lubricants [71, 72, 89, 92, 106-114]. Thus, such scale mitigates the proceeding fretting wear process.

On the contrary, low loads and large amplitudes enhance the release of debris and favour the surface-to-surface contact and the fretting wear of the components.

The non-monotonic relationship (presence of a turning point) between fretting wear and load indicates a change in the fretting process that is most probably related to a change in the fretting regime. To investigate this aspect, in Fig. 6.61 the specific wear coefficient, calculated according to Eq. 4.7, is plotted as function of the applied load for imposed amplitudes of 35 (pink diamonds), 75 (black squares) and 165 μm (blue circles).

According to the aspects discussed so far, fretting in liquid Pb can be considered as a type of fretting occurring at high temperature in lubricating conditions. As a consequence, it is reasonable to assume that the fretting wear regimes existing for fretting in air and lubricated fretting (paragraph 2.4) are applicable also to the case of fretting in molten Pb. For this reason, the curve specific wear coefficient – load described in [63] for fretting in normal atmosphere with fixed imposed sliding amplitude is plotted in Fig. 6.61 (grey line) for comparison. Adapting this curve to fretting in molten Pb with sliding amplitude of 35 μm, it can be noticed that the experimental points seem to follow the trend proposed in [63].

The shift towards lower values of specific wear coefficient is due to the lubricating action of liquid Pb (tackled at the beginning of this paragraph) together with the different testing conditions (e.g. amplitude tested > 90 μm) and materials (only austenitic steels) considered in [63].

The lubricating action exerted by liquid Pb also explains the difference, observable in Fig. 6.61, between the specific wear coefficients calculated for fretting tests performed in liquid Pb (black squares) and in normal atmosphere (empty red squares – adapted from Fig. 5.23) with the same sliding amplitude (i.e. 75 μm).

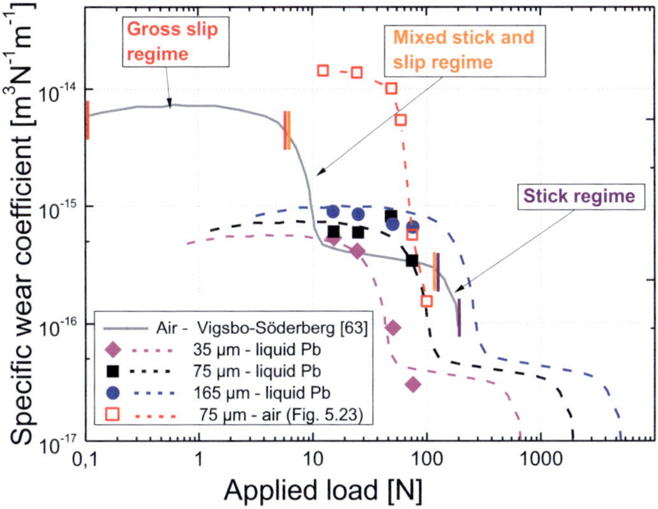

Fig.6.61. Fretting map of specific wear coefficient vs applied load for friction pair of T91 steel. Curves for imposed amplitudes of 35, 75 and 165 µm are plotted. Test parameters: 450°C, 10^{-6} wt% [O], 10 Hz, $5.4 \cdot 10^{6}$ cycles, 150 h.

Adapting the curve of reference (grey line) to the other imposed amplitudes (75 and 165 µm), it can be supposed that a load increase tends to bring the contact conditions towards the stick regime, where the specific wear rate (and the fretting wear) is minimized.

Moreover, for the same applied loads, different fretting regimes might exist for different imposed amplitudes. Indeed, with sliding amplitude of 35 µm, applied loads higher than 25 N seem to shift the system from gross-slip to mixed stick-slip regime. With sliding amplitude of 75 and 165 µm the same shift is expected for loads around 50 N and for loads higher than 75 N respectively. Additionally, as suggested by Fig. 6.61, it is expected that the larger the amplitude, the higher the load corresponding to the beginning of the stick regime. Therefore, also for fretting in molten Pb it seems well-grounded to assume that combination of applied load and sliding amplitude determines the fretting regime.

As shown in Fig.6.62, for friction pairs of 15-15Ti, the same trend as for T91 was noticed at the same testing conditions (450°C for 150 h with sliding amplitude of 75 μm and applied loads between 15 and 75 N).

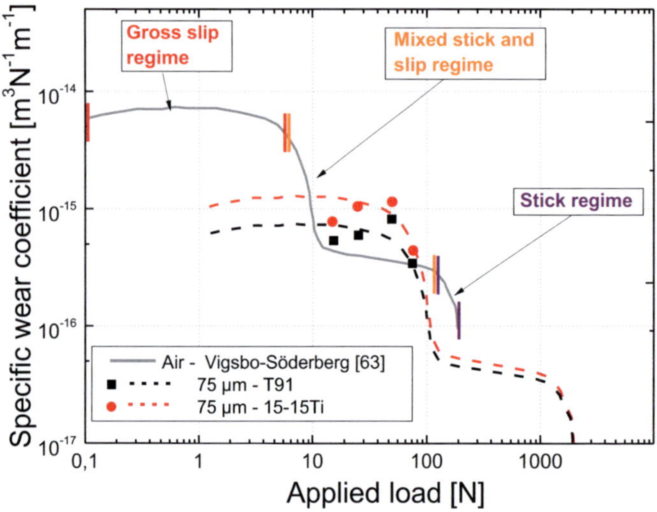

Fig.6.62. Fretting map of specific wear coefficient versus applied load at fixed imposed amplitude of 75 μm for friction pair of T91 steel and 15-15Ti. Test parameters: 450°C, 10^{-6} wt% [O], 10 Hz, $5.4 \cdot 10^{6}$ cycles, 150 h.

The data of 15-15Ti are shifted to higher values of specific wear coefficient because, as previously explained, 15-15Ti is more susceptible to fretting than T91.

As mentioned for Fig. 6.61, the curves in Fig. 6.62 are shifted towards lower values of specific wear coefficient compared to the reference curve described by Vingsbo and Södergberg in [63] because of the lubricating action of Pb and the different testing conditions.

The understanding of the amplitude effect on the fretting process was deepened through tests carried out on all the materials under investigation with constant load (50 N) and amplitude variable between 15 and 190 μm.
A general increase of fretting wear with amplitude was noticed. Such increase, as explained for fretting in air and as suggested by the variation in fretting depth and specific wear coefficient displayed in Fig. 6.35 and 6.37, is not merely due to an increase of the fretting affected area but must be related to change in the fretting regime.
For friction pairs of T91 (black squares), 15-15Ti (red circles) and GESA-T91 (green triangles), the variation of specific wear coefficient with the amplitude is shown in Fig. 6.63 together with the curve specific wear coefficient – amplitude described in [63] for fretting in normal atmosphere (grey line) and fixed applied load.
The susceptibility of the material to fretting and to the corrosion processes occurring in liquid Pb determines the position of the plots in Fig. 6.63. In liquid Pb with 10^{-6} wt% of oxygen at 450°C, GESA-T91 showed the highest wear and oxidation/dissolution resistance, T91 follows thanks to the protecting oxide scale which partially mitigate the fretting wear and 15-15Ti, most likely due to dissolution attacks, occupy the highest position in Fig. 6.63.
The distribution of the experimental data seems to agree with the general trend of the curve reported in [63] (grey line). The presence of liquid Pb (lubrication and corrosion processes) together with the different testing conditions (e.g. applied load < 36 N) and materials (only austenitic steels) treated in [63] are likely the reasons for the lower specific wear coefficients measured in this work compared to the reference curve (grey line).

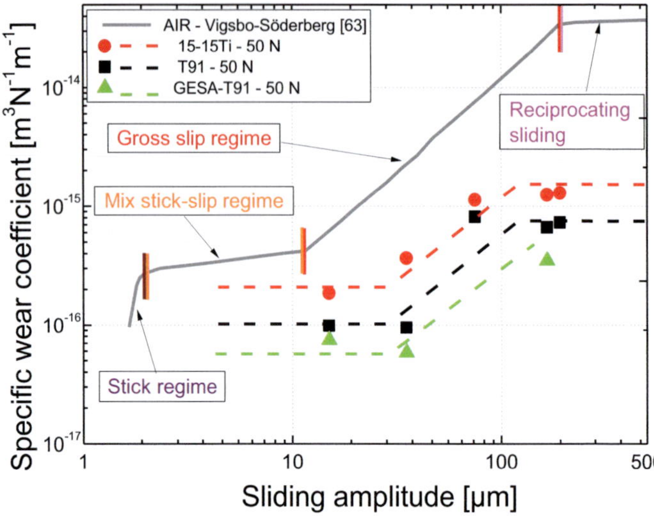

Fig.6.63. Fretting map of specific wear rate vs sliding amplitude with constant applied load of 50 N for friction pairs of T91, 15-15Ti and GESA-T91. Test parameters: 450°C, 10^{-6} wt% [O], 10 Hz, $5.4 \cdot 10^{6}$ cycles, 150 h.

For amplitudes shorter than 35 µm, for all the materials under investigation the specific wear coefficient does not change significantly whereas a significant transition occurs for larger amplitudes. This indicates most likely a change of fretting regime roughly around 35 µm which, according to the reference curve [63], is supposed to be from gross slip to mixed stick-slip regime. Thus, the sudden increase of fretting wear, for amplitudes larger than 35 µm (Figs. 6.24 – 6.27), is most likely related to a change of fretting regime.

Besides, the additional data points available for T91 and 15-15Ti at 190 µm are characterized by a specific wear coefficient very similar to the one measured for a sliding amplitude of 165 µm. This observation, according to the reference curve [63], might indicate a further change in fretting regime

around 165 μm that is most likely from gross slip regime to reciprocating sliding.
The interpretation of the experimental outcomes plotted in Fig. 6.63, was made also on the basis of the indications given by Fig. 6.61 and 6.62. In these figures, an applied load of 50 N is expected to bring the system in a fretting regime of gross slip when the sliding amplitude is 75 μm and of mixed stick-slip when the amplitude is 35 μm. Besides, Figs. 6.61 and 6.62 suggest that in liquid Pb the stick regime might be expected for a load of few hundred N and that the larger the amplitude the higher this threshold. Thus, with reference to Fig. 6.63 and taking into account also the lubricating action of molten Pb, it can be hypothesized that with an applied load of 50 N, the stick regime might be reached only with very short amplitudes, likely of few microns.
The presence of liquid Pb, besides shifting the stick regime towards lower amplitudes (Fig. 6.63) and higher loads (Fig. 6.61 and 6.62), is expected to affect the transition between gross slip and mixed stick-slip regime and their extension. The experimental results showed that the mixed stick-slip regime seems enlarged compared to the case in air (to the detriment of gross slip regime if the reciprocating sliding regime is considered to be shifted to shorter amplitude due to the lubrication by liquid Pb). The enlargement of the mixed stick-slip regime is most likely favoured by the higher debris retention occurring in Pb than in air. The much higher density of Pb than the one of air enhances debris entrapment in the contact area. Under local pressure created by the applied load, the compaction of metal debris, oxidized debris and liquid Pb at 450°C might favour the formation of micro-welding points between specimens and counter-specimens. Moreover, dissolution of alloying elements (e.g. Ni) in liquid Pb enclosed in the contact area, can causes a change of composition that might locally induce solidification.

A graph load – amplitude, which contains the border lines of the fretting regime corresponding to certain combinations load/amplitude in liquid Pb, can be created as shown in Fig. 6.64.

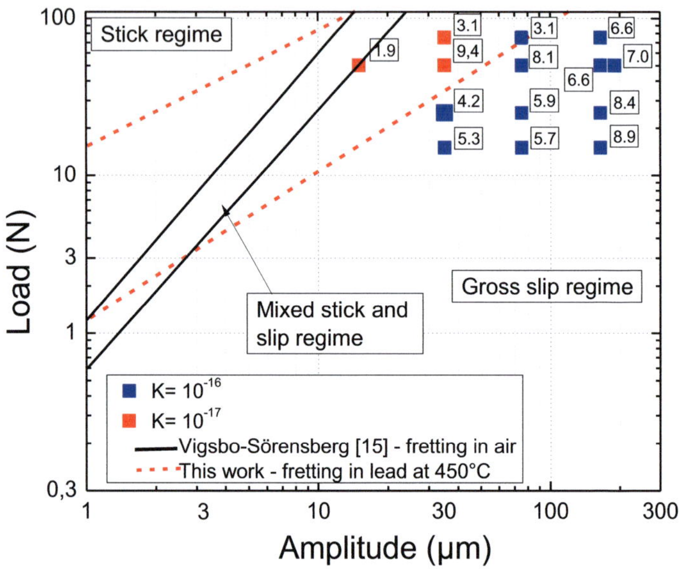

Fig.6.64. Fretting map sliding amplitude vs applied load for T91. Each squared point in the graph correspond to a different load/amplitude combination at 450°C, with 10^{-6} wt% [O], after $5.6 \cdot 10^{6}$ cycles / 150 h.

In Fig 6.64, data points represent load-amplitude combinations experimentally tested and their colour reflect the order of magnitude of the corresponding specific wear coefficient. Although experimental data are not sufficient to draw precise fretting regime border lines, the expected regime domains can be suggested. The considerations made on the basis of Figs. 6.61, 6.62 and 6.63, suggested that: at 450°C fretting regimes in liquid Pb are expected to be shifted towards shorter amplitudes and higher loads; the mixed stick-slip regime might be enlarged due to the enhanced debris retention of fretting in Pb and, as consequence, the gross-slip regime likely shrinks; the combinations 75 μm / 50-75 N and 35 μm / 50 N are most likely at the transition between mixed stick-slip regime and gross slip regime. As a result, in Fig. 6.64, the border lines of the fretting regime proposed by Vingsbo and Söderberg in [63] for fretting in air (black line) might be shifted according to the red dashed lines.

The influence of combinations amplitude/load on the fretting process can be represented also in 3D fretting maps like the ones shown in Figs. 6.65, 6.66, 6.67 and 6.68. In these graphs, the x and y-axis are load and amplitude respectively, while for the z-axis different parameters like fretting depth, penetration rate or specific wear coefficient can be used to express the fretting wear severity. These 3D maps were constituted starting from the experimental results and adding to them expected values of specific wear coefficients that are qualitatively predicted according to the expected fretting regimes for friction pairs of T91, 15-15Ti and GESA-T91 (Figs. 6.61 – 6.64). In this way, a wider overview of the wear process can be provided. In the 3D fretting maps, full circles correspond to values experimentally measured; empty circles represent the expected values according to the fretting regimes plot (Figs. 6.61 – 6.64) interpretation.
Fig. 6.65 and 6.66 contain fretting maps for friction pairs of T91 relative to maximum fretting depth and penetration rate (μm/h) respectively. In these fretting maps, it can be clearly noticed that for short amplitudes (e.g. < 35 μm) and high loads (e.g. > 50 N) the fretting damage is minimized. However, in the case of unavoidable large amplitudes (> 75 μm), low loads (< 25 N) can also reduce the fretting damage.

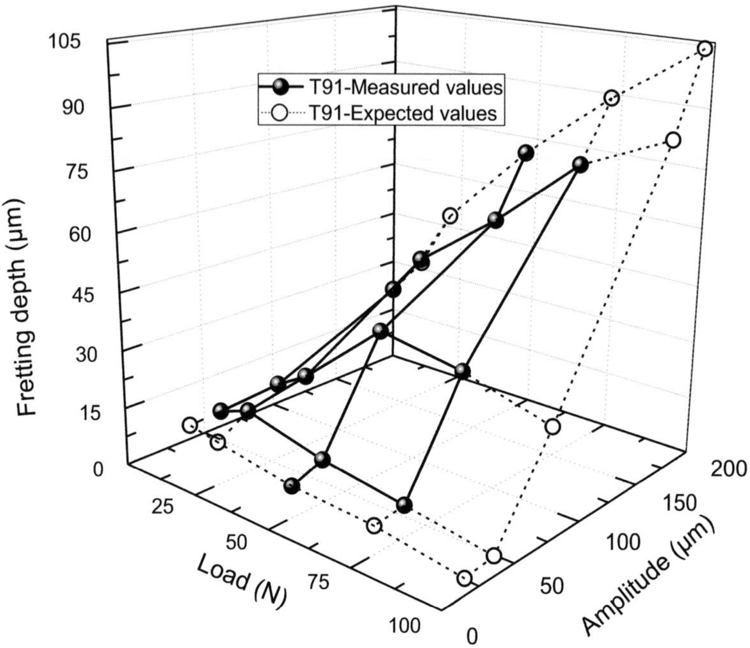

Fig.6.65. 3D fretting map for friction pairs of T91. The maximum fretting depth (z axis) is plotted versus load (axis-x) – amplitude (axis-y) combinations. Test parameters: 450°C, 10^{-6}wt% [O], 10 Hz, $5.6{\cdot}10^{6}$cycles, 150 h.

Fig. 6.67 results from the 3D combination of Fig. 6.61 and 6.63 and it shows a 3D fretting map in terms of specific wear coefficient for friction pairs of T91.

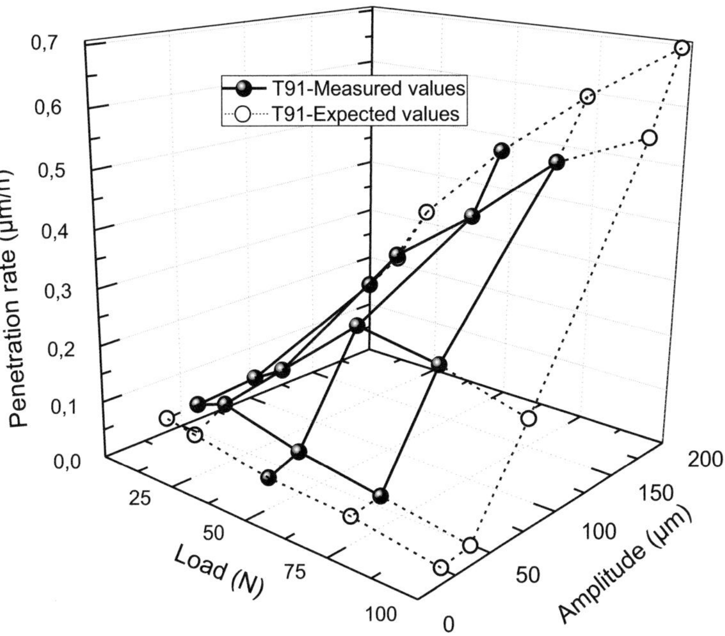

Fig.6.66. 3D fretting map for friction pairs of T91. The penetration rate (z axis) is plotted versus load (axis-x) – amplitude (axis-y) combinations. Test parameters: 450°C, 10^{-6} wt% [O], 10 Hz, $5.6\cdot10^{6}$ cycles, 150 h.

This 3D map underlines that less severe fretting conditions are established when high applied loads are combined with short sliding amplitudes. This combination of variables brings the system in the mixed stick-slip regime (and towards the stick regime) where the fretting wear is lower than in the gross slip regime. However, in the mixed stick-slip regime other fretting related issues (fretting fatigue) might become predominant (paragraph 2.4).

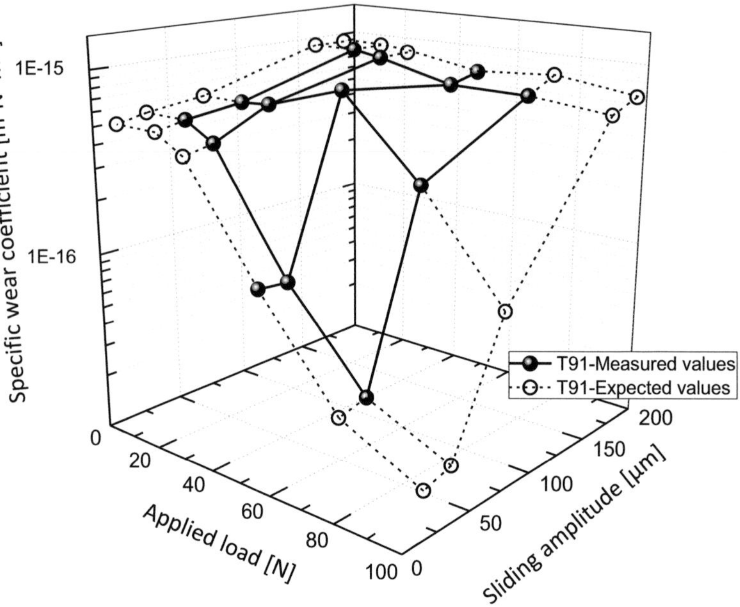

Fig.6.67. 3D fretting map for friction pairs of T91. The specific wear rate (z axis) is plotted versus load (axis-x) – amplitude (axis-y) combinations. Test parameters: 450°C, 10^{-6} wt% [O], 10 Hz, $5.6 \cdot 10^6$ cycles, 150 h.

In the 3D fretting map of Fig. 6.68, the specific wear coefficient values for friction pairs of 15-15 Ti (red circles) and GESA-T91 (green circles) are compared to the ones of T91 (black circles).

Due to the higher fretting and corrosion susceptibility, the experimental points for 15-15Ti are shifted towards higher values of specific wear coefficient. On the contrary, the high wear and corrosion resistance of GESA-T91 leads to a shift towards lower values of the measured specific wear coefficient.

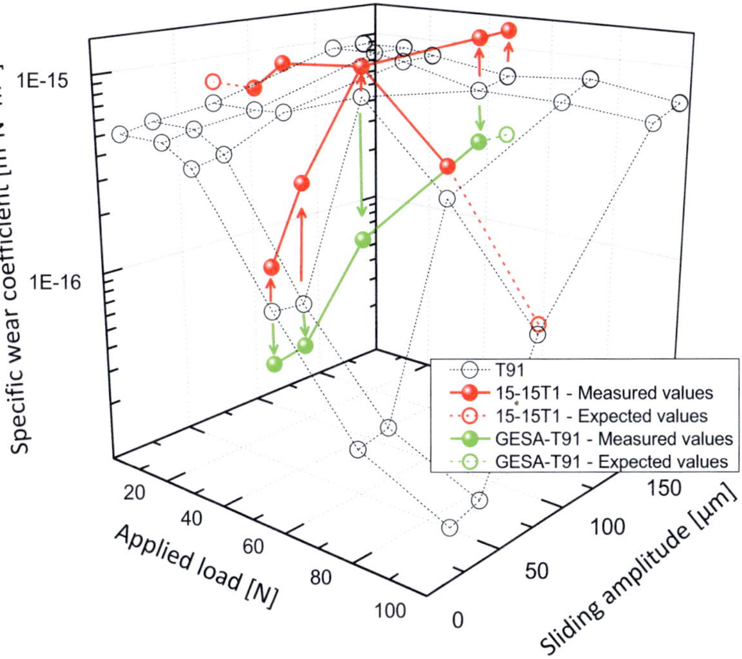

Fig.6.68. 3D fretting map for friction pairs of 15-15Ti, T91 and GESA-T91 The specific wear rate (z axis) is plotted versus load (axis-x) – amplitude (axis-y) combinations. Test parameters: 450°C, 10^{-6} wt% [O], 10 Hz, $5.6 \cdot 10^{6}$ cycles, 150 h.

6.2.6 Temperature effects

The experimental outcomes underline the fundamental role of the temperature in the fretting process and they highlight that the temperature has a remarkable influence on the fretting behaviour of the investigated alloys already after 150 hours ($5.4 \cdot 10^{6}$ cycles). As shown in Fig. 6.69, increasing the temperature from 450 to 550°C, the fretting depth increases by a factor 4, 2.7 and 1.4 for T91, 15-15Ti and GESA-T91 respectively.

The temperature increase enhances significantly the fretting wear, especially for T91 and 15-15Ti, which at 550°C, already after 150 h, are affected by 30 and 22 % of fretting penetration respectively. GESA-T91 is far less susceptible to the temperature increase and the fretting penetration is only 4 %.

For all the tested materials, an exponential law well fits the evolution with the temperature of fretting depth and volume loss. Additionally, it can be observed that for temperature above 500°C the T91 curve overcomes the one of 15-15Ti and that the GESA-T91 curve is far lower than the other two.

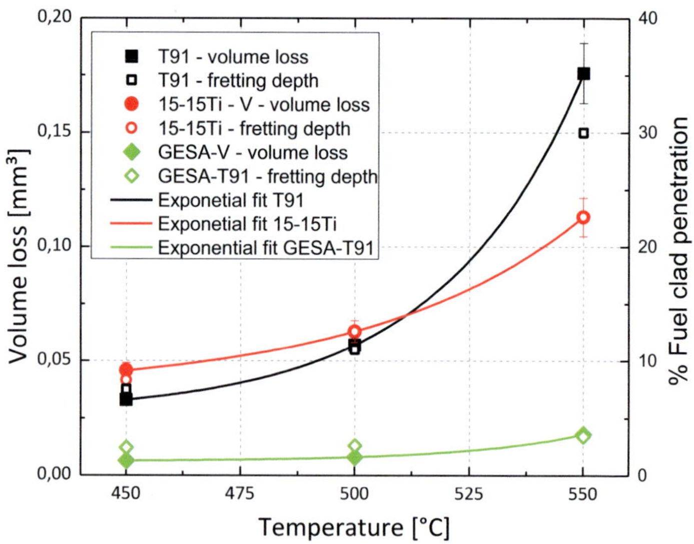

Fig.6.69. Graph of volume loss and fretting depth versus temperature for friction pairs of T91, 15-15Ti and GESA-T91. Test parameters: 50 N, 75 µm, 10 Hz, 10^{-6} wt% [O], $5.4 \cdot 10^{6}$ cycles, 150 h.

The observations made so far can be explained considering that the corrosion processes occurring in liquid Pb (oxidation and dissolution) are diffusion dependent and hence temperature dependent according to an Arrhenius type relation. This most probably explains the exponential behaviour of the graphs drawn in Fig. 6.69. Additionally, the significant increase of the T91 curve can be associated with the extensive oxidation (oxidation enhanced fretting) affecting this steel at 550°C with 10^{-6} wt% of oxygen [2, 39-55]. Composition and martensitic structure of T91 steel together with the high temperature (550°C) favour inwards oxygen diffusion and outward metal cations diffusion. As a result, T91 steel is affected by extensive oxidation characterised by the formation of a thick duplex oxide scale and of an internal oxidation zone (see Fig. 6.44). Although in the first stage of fretting the fast growing oxide scale might favour a sort of lubricating action, in the long run it likely enhances the material removal. Indeed, oxide scales are less ductile than the original alloy and the thicker is the oxide scale the more considerable is this aspect. Moreover, thick oxide scales are generally unstable and tend to break off. On the contrary, composition and austenitic microstructure of 15-15Ti steel, as previously described, avoid the extensive oxidation (at the tested temperature and oxygen content). Up to 500°C, the dissolution affecting 15-15Ti and the less severe oxidation of T91 are likely the reason behind the higher fretting susceptibility of 15-15Ti.

The alumina scale forming on GESA-T91, characterized by a slow growing rate, high adhesion, reduced thickness and the good mechanical properties, ensures protection from dissolution and extensive oxidation resistance together with good wear resistance. As a result, the temperature influence is limited and the penetration is kept below 5 % even at 550°C. For these peculiarities, at higher temperature, the gap between GESA-T91 and the other materials under investigation in this work becomes more evident.

With reference to the 3D fretting maps of Fig. 6.67 and 6.68, according the experimental outcomes, for all the materials, a temperature increase is expected to cause a shift towards higher values of specific wear coefficient (Fig. 6.70).

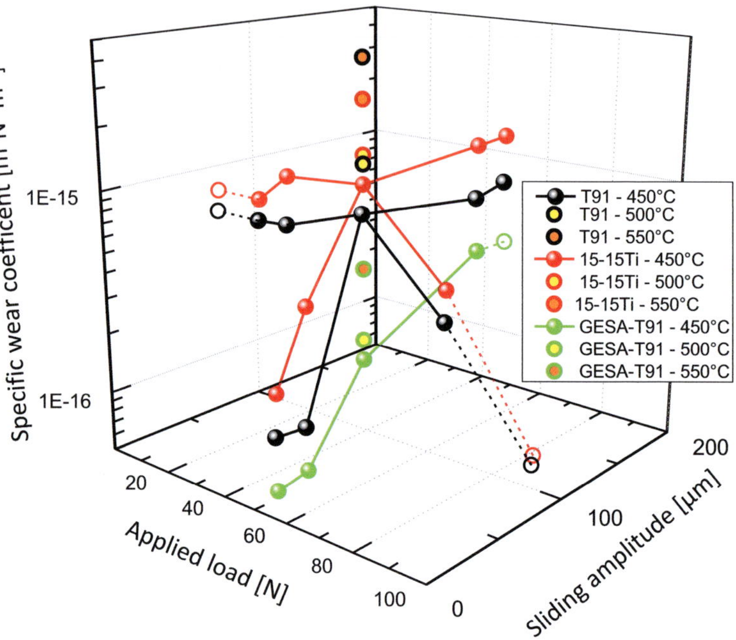

Fig.6.70. 3D fretting map for friction pairs of T91, 15-15Ti and GESA-T91. The specific wear rate (z axis) is plotted as function of the load (axis-x) – amplitude (axis-y) combination. The shift of the data determined by the temperature increases is also represented. Test parameters: 10^{-6} wt% [O], 10 Hz, $5.6 \cdot 10^6$ cycles, 150 h.

6.2.7 Final comparison

To compare and represent in one single graph the results deriving from different fretting tests performed in liquid Pb, the plot wear rate versus work rate is built in Fig. 6.71 according to Eq. 5.1. This figure offers an overview of the influence of the different tested parameters on the fretting process after 150 h and underlines that, among the tested conditions, the lowest wear rates corresponds to combinations of high load (≥ 50 N) and short sliding amplitude (≤ 35 µm).

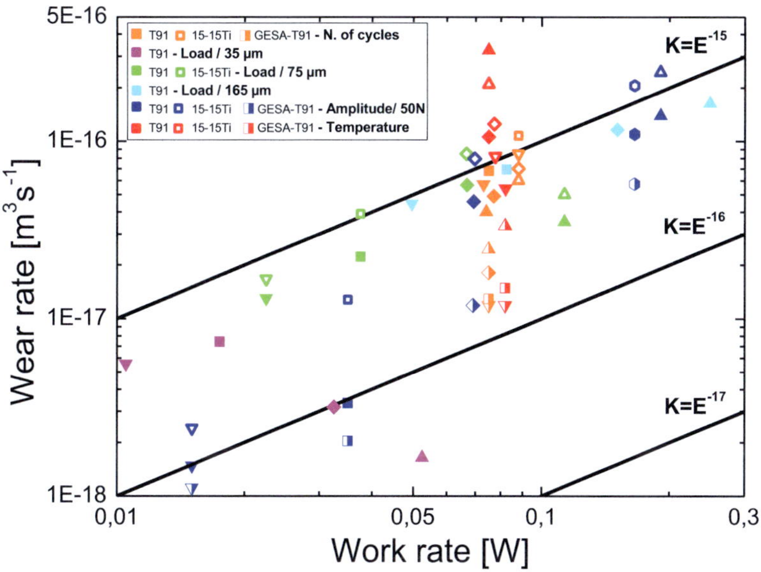

Fig.6.71. Graph wear rate vs work rate for different fretting conditions in molten Pb for T91, 15-15Ti and GESA-T91.

As an alternative comparison approach, the penetration rate can be plotted versus the work rate. Such graph, shown in Fig. 6.72, can be used to predict fretting penetration after a certain time.

Considering 10 % as the maximum allowed fretting penetration of the fuel clad (arbitrary selection), about 25000 h (3 years) as the targeted operating lifetime in the reactor for fuel elements and a working temperature between

400 and 480°C (see Tab. 1.2), a graph like the one of Fig. 6.72 can indicate which combination of fretting parameters meet these requirements (e.g. a penetration rate around 0.002 μm/h).

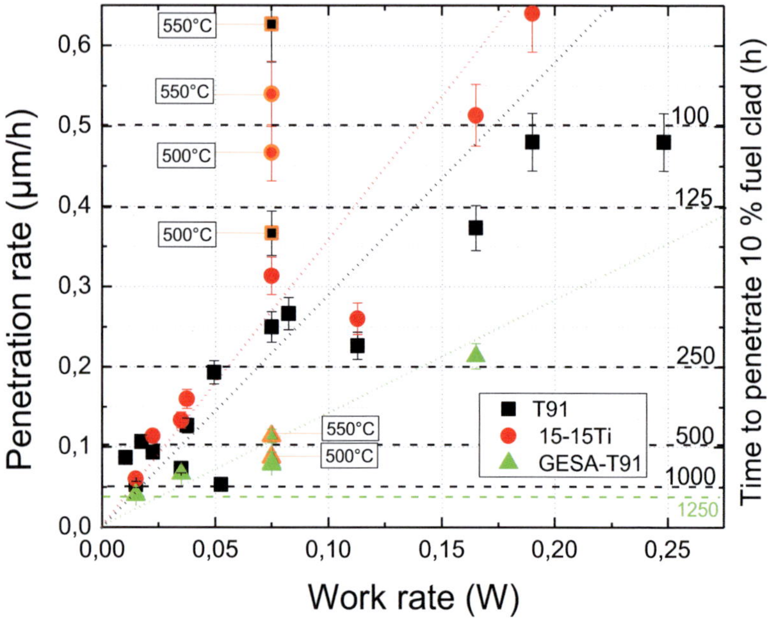

Fig.6.72. Graph penetration rate vs work rate for different fretting conditions in molten Pb for T91, 15-15Ti and GESA-T91.

Fig. 6.72 points out that the fretting conditions experimentally tested lead to penetration rates unacceptable for real applications. Even for the lowest measured penetration rates (resulting from an applied load of 50 N and a sliding amplitude of 15 μm), which for 15-15Ti, T91 and GESA-T91 are 0.063, 0.053 and 0.039 μm/h respectively, the penetration limit would be reached after about 800, 950 and 1300 h. Such timing does not meet the targeted operating lifetime for fuel elements, which is of about 25000 h.

However, the data of Fig. 6.72 refer to fretting tests carried out in rather severe conditions both in terms of load-amplitude combinations (typical of the mix stick-slip regime) and exposure time (150 h / $5.4 \cdot 10^6$ cycles is within the run-in period, when fretting is severe at the most).
The fretting tests carried out in this work and used to compile Fig. 6.72, can be interpreted as a kind of accelerated test that are useful to define the fretting regimes, to extrapolate data for predicting the long time fretting behavior and to provide to the designers indications to prevent a severe fretting damage.
A close analysis of the experimental outcomes indicates that it is possible to comply with the mentioned operating requirements by considering the following aspects:

- Favorable load-amplitude combinations can minimize the fretting severity. The fretting penetration and the wear rate decrease with the increasing applied load (after the turning point) and with the decreasing sliding amplitude (paragraphs 6.1.3 and 6.1.4). As displayed by the fretting maps of Figs. 6.61-6.67, there are load-amplitude combinations (e.g. load > 75 N and amplitude < 15 μm) that favors the evolution of the fretting contact towards the stick regime. These load-amplitude combinations are expected to lower the specific wear coefficient (and consequently the wear and penetration rate) by up to 1 order of magnitude compare to e.g. the combination 50 N-15 μm, which in Fig. 6.72 corresponds to the lowest measured penetration rates.
- Such conditions can be established with an adequate assembly strategy.

- Suitable material improvement and corrosion counter-measures improve the wear resistance and mitigate the corrosion enhancement of fretting.
- A reduction of the penetration rate up to about 50 % already after 150 h was achieved by using pre-oxidized components (Fig. 6.47). Likewise, for the specific case of austenitic steels, the use of Ni-enriched Pb reduces the fretting corrosion severity (e.g. Fig. 6.52).
- Additionally, surface treatment and Al-alloying (e.g. GESA-T91) improves remarkably the wear and the corrosion resistance property also at high temperature (e.g. Fig. 6.60 and 6.69).
- Fretting is a self-mitigating process. Fretting becomes less and less severe with the time (Figs. 6.12-6.15). After 930 h, the penetration rate is up to 70% lower than during the run-in period (Fig. 6.13) to which the data of Fig. 6.72 refer.
- The beginning of the fretting process is delayed with respect of the reactor start-up. In e.g. the reactor core, the fretting process starts when, due to the irradiation damage and thermal-mechanical loads, the components (e.g. springs, supports or spacer grids) aimed to fasten the fuel cladding tubes (or other parts) lose their mechanical and dimensional properties. Therefore, the fretting process affect the e.g. fuel cladding tube for less than 25000 h.
- Possible design solution like the installation of spacer grids and supports in the non-active area of the fuel (lower temperature and irradiation) can delay further the beginning of the fretting process.

It is possible to claim with a certain confidence that, the listed aspects will allow the achievement of the penetration rate complying with the operating requirement (0.002 µm/h), which is about 1 order of magnitude lower than the one measured experimentally (in the best cases from 0.039 to 0.063 µm/h for the three materials).

According to the indications emerging from this chapter, to comply with the operating conditions, the designers should foresee assembly strategies able to keep the sliding amplitude lower than 10 µm, to reduce as much as possible the components vibration frequencies (smaller penetration for constant exposing time) acting e.g. on the coolant flow and velocity and to impose fastening loads higher than 100 N. Finally, to face fretting and fretting corrosion at the best, the use of surface aluminized steels, like the investigated GESA-T91, should be considered.

7. Summary and conclusion

In the present work, for the first time the fretting process of different steels was investigated in molten lead, focusing on its interaction with the corrosion mechanisms affecting steels in this medium. This target was reached through the successful design and realization of an original, dedicated and reliable facility, the FRETHME facility, equipped to run fretting tests in liquid Pb/Pb alloys in nuclear reactor relevant conditions.

The preliminary experimental campaign carried out in normal atmosphere on T91 steel, showed good agreement with the literature database, indicating an adequate investigating approach of the fretting process, an appropriate design of the facility and of the parameter control system and a correct execution of the fretting tests.

The results of the research activity centered on fretting in molten Pb provided several important information that will be required for designing purposes of future lead cooled nuclear systems.

On the one hand, fretting influences the corrosion behavior in molten Pb of candidate steel for nuclear application. Protective oxide scales and corrosion barriers, necessary to hinder diffusion and extensive oxidation affecting steels in liquid lead, are gradually removed by the fretting action. As a result, liquid Pb reaches the bare steel surface and severe corrosion can take place. For example, due to the fretting actions, 15-15Ti is affected by dissolution at lower temperature and after shorter exposure time than without fretting.

On the other hand, the corrosion mechanisms can enhance the fretting process. From the experimental outcomes it emerges that higher corrosion susceptibility results in higher fretting damage. Up to 500°C, for every tested condition, 15-15Ti steel showed the most significant fretting damage due to the dissolution process occurring in the fretted area (dissolution enhanced fretting). At 550°C, the extensive oxidation of T91 boosts the fretting process more than the dissolution so that at this temperature, T91 is more severely damaged by fretting than 15-15Ti (oxidation enhanced fretting). For the same reasons the surface aluminized T91 (GESA-T91), characterized by the best corrosion properties, showed the highest fretting wear resistance. Moreover, the mitigation of the corrosion processes, by e.g. the use of Ni-enriched Pb or of pre-oxidized friction pairs, can significantly reduce the fretting wear.

Despite of the corrosion enhancement, fretting in liquid lead is less severe than in air because of the lubricating action of lead.

The fretting affecting parameters determine the severity of fretting wear in liquid lead. Fretting is a self-mitigating process, characterized by a progressive reduction of the wear rate with increasing time / number of cycles. Temperature increase, due to its influence on the corrosion processes occurring in liquid lead, enhances fretting wear. Combinations of load and amplitude strongly influence the fretting regime and the deriving wear severity. The experimental campaign pointed out that the fretting wear can be minimized by bringing the system in the stick regime that is favored by a combination of short amplitude (e.g. < 10 µm) and high load (e.g. > 100 N).

Among the materials investigated, the GESA treatment of FeCrAlY plasma sprayed T91 (GESA-T91) gives the best guarantees in terms of fretting wear resistance together with optimal corrosion properties. At every tested condition this material showed the best behavior and the highest stability (even at elevated temperature) among the alloys under study. T91 up to 500°C shows generally a better behavior than 15-15Ti, while at higher temperature the converse is true. For most of the tested conditions both of these materials show an unacceptably severe fretting wear with respect to real applications. However, when fretting shifts closer and closer to the stick regime, the gap between these alloys and the GESA-T91 becomes smaller.

The experimental campaign returns a set of data that were used to constitute preliminary fretting maps useful for fretting damage predictions. Such maps represent a powerful tool to evaluate possible fretting scenarios and determine the conditions necessary to operate within the safety margins. Assuming certain safety parameters (i.e. maximum fretting depth and time to reach it), graphs similar to Fig. 6.72 indicate to designers which range of fretting parameters, i.e. applied load and sliding amplitude, must be established to operate within the safety margins. This is expected to play a significant role in the design procedure like e.g. the assembly strategies for the reactor components.

In this work, 10 % of the total fuel cladding tube thickness was arbitrarily assumed as maximum acceptable damage. Such penetration depth at 450°C was reached with the lowest tested penetration rate (resulting from sliding amplitude of 15 μm and applied load of 50 N) after 1300, 950 and 800 h for GESA-T91, T91 and 15-15Ti respectively. Such timing does not meet the targeted operating lifetime for fuel elements, which is of about 25000 h. However, a close analysis of the experimental outcomes and the extrapolation of data for long term fretting exposure indicate that it is possible to comply with the mentioned operating requirements by establishing a favorable combination of fretting parameters and adopting suitable countermeasures to mitigate the corrosion enhancement of fretting.

This work provides a description of fretting in liquid Pb at high temperature and a general understanding of the possible approaches to minimize its effects on the reactor components lifetime. Although it was not possible to perform fretting tests with durations comparable to the reactor core components lifetime, the well understanding of the influence of the fretting variables (e.g. temperature, number of cycles, applied load and sliding amplitude) and of the fretting process evolution in the time allow to use the experimental data of this work to propose long term predictions of fretting behavior of the investigated materials.

With respect to real applications, this work pointed out the non-negligibility of the fretting process for essential nuclear reactor components like fuel cladding tubes or steam generator tubes. As demonstrated by the experimental campaign, an uncontrolled fretting process can cause severe damage and eventually component failures that represent key issues for safety and economy aspects.

References

[1] L. Cinotti, C. F. Smith, J. J. Sienicki, H. Ait Abderrahim, G. Benamati, G. Locatelli, S. Monti, H. Wider, D. Struwe, A. Orden, I.S. Hwang*et*, The potential of LRF and ELSY project, Proceedings of ICAPP 2007, Nice, France, May 13-18.

[2] NEA (Nuclear Energy Agency), Handbook on Lead-bismuth Eutectic Alloy and Lead Properties, Materials Compatibility, Thermal-hydraulics and Technologies, OECD Nuclear Science, 2007.

[3] NEA (Nuclear Energy Agency), Accelerator-drive Systems (ADS) and Fast Reactors (FR) in Advanced Nuclear Fuel Cycles – A comparative study, 2002.

[4] C.F. Smith, I.G. Halsey, N.W. Brown, J.J. Sienicki, A. Moisseytsev, D.C. Wade, SSTAR: the US lead-cooled fast reactor (LFR), Journal of Nuclear Materials, 376 (2008) 255-259.

[5] S. Glasstone, A. Sesonke, Nuclear Reactor Engineering, Chapman & Hall, New York, USA, 1994.

[6] D. Gorse-Pomonti, V. Russier, Liquid metals for nuclear applications, Journal of non-Crystaslline Solids, 353 (2007) 3600-3614.

[7] P. Hejzlar, N.E. Todreasa, E. Shwageraus, A. Nikiforovaa, R. Petroskia, M.J. Driscoll, Cross-comparison of fast reactor concepts with

various coolants, Nuclear Engineering and Design, 239 (2009) 2672-2691.

[8] K. Tuček, J. Carlsson, H. Wider, Comparison of sodium and lead-cooled fast reactors regarding physics aspects, severe safety and economical issues, Nuclear Engineering and Design, 236 (2006) 1589-1598.

[9] L. Cinotti, C.F. Smith, H. Sekimoto, L. Mansani, M. Reale, J.J. Sienicki, Lead cooled system design and challenges in the frame of Generation IV International Forum, Journal of Nuclear Materials, 415 (2011) 245-253.

[10] B.F. Gromov *et al.*,The problem of technology of the liquid metals coolants (lead-bismuth, lead), Stare scientific centre of Russian Federation, Institute of Physics and Power Engineering (IPPE), Obninsk, Russia.

[11] B.F. Gromov, Designing the Reactor Installation with Lead-Bismuth Coolant for Nuclear Submarines. The Brief History. Summarised Operations Results, Proceedings of the Conference Heavy Liquid Metals Coolants in Nuclear Technologies (HCLM-98), Vol. 1, 14, Obninsk, 1998.

[12] T.R. Allen, D.C. Crawford, Lead-Cooled Fast Reactor Systems and the Fuel and Materials Challenges-Review Article, Science and Technology of Nuclear Installation, Volume 2007, Article ID 97486, 11 pages.

[13] G. Müller, G. Schumacher, F. Zimmermann, Investigation on controlled liquid lead corrosion of surface treated steels, Journal of Nuclear Materials, 278 (2000) 85-95.

[14] A. Weisenburger, A. Heinzel, G. Müller, H. Muscher, A. Rousanov, T91 cladding tubes with and without modified FeCrAlY coatings

exposed in LBE at different flow, stress and temperature conditions, Journal of Nuclear Materials, 276 (2008) 274-281.

[15] Proceedings of the International DEMETRA workshop on Development and Assessment of Structural Materials and Heavy Liquid Metal Technologies for Transmutation Systems, Journal of Nuclear Materials, 415 (2011) 227-385.

[16] [16] A. Alamberti *et al.*, European lead-cooled fast reactor, FISA 2009, Prague, Czech Republic, 22-24 June,2009.

[17] [17] C.F. Smith, L. Cinotti, H. Sekimoto, Lead-cooled Fast Reactor (LFR) ongoing R&D and key issues, Proceedings of the GIF Symposium 2009, Paris, France, 9-10 September,2009.

[18] [18] L. Cinotti, Design for LFR and ADS, VELLA Thematic School, Karlsruhe, Germany, November 12-14, 2008.

[19] [19] L. Cinotti, LFR Lead-Cooled Fast Reactor, FISA 2006 , Kirchberg, Luxemburg, May 13-16, 2006

[20] [20] A. Gulevich, V. Chekunov, O. Fokina, O. Komlev, O. Kukharchuk, C. Melnikov, N. Novikova, L. Ponomarev, E. Zemskov, Concept of electron accelerator-driven system based on sub-critical cascade reactor, Progress in Nuclear Energy, 50 (2008) 347-352.

[21] [21] F. Groeschel *et al.*, The MEGAPIE 1 MW target in support to ADS development: status of R&D and design, Journal of Nuclear Materials, 335 (2004) 156-162.

[22] [22] C. Rubbia *et al.*, Conceptual Design of a Fast Neutron Operated High Power Energy Amplifier, CERN/AT/95-44 (ET), 29 September 1995.

[23] [23] J.U. Knebel, X. Cheng, C.H. Lefhalm, G. Müller, G. Schumacher, J. Konys, H. Glasbrenner, Design and corrosion study of a

closed spallation target module of an accelerator-driven system (ADS), Nuclear Engineering and Design, 202 (2000) 279-296.

[24] A. Herrera-*Martinez et al.*, Transmutation of nuclear waste in accelerator-driven systems: thermal spectrum, Annals of Nuclear Energy, 34 (2007) 550-563.

[25] A. Herrera-*Martinez,* Y. Kady, G.T. Parks, Transmutation of nuclear waste in accelerator-driven systems: fast spectrum, Annals of Nuclear Energy, 34 (2007) 564-578.

[26] Y. Kadi, Design of an Accelerator-Driven System for the Destruction of Nuclear Waste, Workshop on Hybrid Nuclear Systems for Energy Production, Utilisation of Actinides & Transmutation of Long-Lived Radioactive Waste, Trieste, 3 -7 September, 2001.

[27] A. Berbensi, G. Corsini, L. Mansani, C. Artioli, G. Glinatsis, *EFIT*: the European Facility for Industrial Trasmutation of Minor Actinides.

[28] F. Roelofs, B. de Jager, A. Class, H. Jeanmart, P. Schuurmans, A. Ciampichetti, G. Gerbeth, R. Stieglitz, C. Fazio, European research on HLM thermal-hydraulics for ADS applications, Journal of Nuclear Material, 376 (2008) 401-404.

[29] Ansaldo Nucleare, Materials and Design for EFIT and XT-ADS.

[30] P. Baeten, MYRRHA design and materials, MATRE-I summer school presentation, Rochehaut s/Semois, October 18-23, 2009.

[31] T.R. Allen, J.T. Busby, R.L. Klueh, S.A. Maloy, M.B. Toloczko, Cladding and duct materials for advanced nuclear recycle reactors, GNEP, JOM, January 2008.

[32] J.W. Bennett, K.A. Horton, Materials requirements for liquid metal fast breeder reactors, Metal Transaction A, 9 (1978) 143.

[33] J. Zhang, N. Li, Oxidation mechanism of steel in liquid-lead alloys, Oxidation of Metals, 63 (2005) 353-381.

[34] J.R. Weeks, Compatibility of Structural Materials with Liquid Lead-bismuth and Mercury, Proceedings of the Symposium on Materials for Spallation Neutron Sources, Orlando, 1997.

[35] G. Mueller, A. Heinzel, G. Schumacher, A. Weisenburger, Control of oxygen concentration in liquid lead and lead – bismuth, Journal of Nuclear Materials, 321 (2003) 256-262.

[36] P.F. Tortorelli, Metals Handbook, Corrosion, 9th Edition, Vol.13.

[37] G. Müller *et al.*, Study on Pb/Bi corrosion of structural and fuel cladding materials for nuclear applications – Part 1: corrosion investigation of steels after 800 and 2000 h exposure in stagnant liquid Pb/Bi at elevated temperatures, JNC TY9400 2002-016.

[38] G. Müller, G. Schumacher, F. Zimmermann, Investigation on controlled liquid lead corrosion of surface treated steels, Journal of Nuclear Materials, 278 (2000) 85-95.

[39] J. Zhang, N. Li, Analysis on liquid metal corrosion-oxidation interactions, Corrosion Science, 49 (2007) 4154-4184.

[40] V. Gorynin, Structural Materials for Power Plants with Heavy Liquid Metals as Coolants, Proceedings of the Conference Heavy Liquid Metals Coolants in Nuclear Technologies(HCLM-98), Vol. 1, 120, Obninsk, 1998.

[41] G. Müller *et al.*, Results of steel corrosion tests on flowing liquid Pb/Bi at 420-600°C after 2000 h, Journal of Nuclear Materials, 301 (2002) 40-46.

[42] F.H. Stott, The protective action of oxide scales in gaseous environments at high temperature, Reports on Progress in Physics., 50 (1987) 861-913.

[43] P. Kofstad, High Temperature Corrosion, Elsevier Applied Science Publishers, Ltd (1987) 413.

[44] D. Struwe, W. Pfrang, Karlsruhe Institute for Technology (KIT), Private Comunication.

[45] J. Zhang, A review of steel corrosion by liquid lead and lead-bismuth, Corrosion Science, 51 (2009) 1207-1227.

[46] L. Martinelli, T. Dufrenoy, K. Jaakou, A. Rusanov, F. Balbaud-Celerier, High temperature oxidation of Fe-9Cr-1Mo steel in stagnant liquid lead-bismuth at several temperatures and for different lead content in liquid alloy, Journal of Nuclear Materials, 376 (2008) 282-288.

[47] M. McLean, K. Sridharan, N. Li, S. Ukai, T. Allen , Time dependence of corrosion in steels for use in lead-alloy cooled reactors, Journal of Nuclear Materials, 371 (2007) 134-144.

[48] O. Yeliseyeva, V. Tsissar, G. Benamati, Influence of temperature on interaction mode of T91 and AISI 316L steels with Pb-Bi melt saturated by oxygen, Corrosion Science, 50 (2008) 1672-1683.

[49] J.M. Herbelin, Effect of Al Addition and Minor Elements on Oxidation Behavior of Fe-Cr Alloys, Journal de Physique IV, 5 (1995) 365-374.

[50] F. Balbau-Celerier, Corrosion of Steels in Liquid Lead Alloys protected by an Oxyde Layer, Application to the MEGAPIE Target and to the Russian Reactor Concept Breast 300, CEA Report, 2003.

[51] J. Zhang, N. Li, Review of Studies on Fundamental Issues in LBE Corrosion, Progress in Material Science, 2004.

[52] G. Mueller *et al.*, Study on Pb/Bi corrosion of structural and fuel cladding materials for nuclear application, Report 1, FZK/IHM, Nuclear Safety Project (NUKLEAR), Karlsruhe, May 2002.

[53] A. Weisenburger *et al.*, Long term corrosion on T91 and AISI 316L steel in flowing lead alloy and corrosion protection barrier development: Experiments and Models, Journal of Nuclear Materials, 415 (2011) 260-269.

[54] A. Weisenburger, A. Jianu, H. Heinzel, M. Del Giacco, R. Fetzer, G. Müller, Compatibility of austenitic steels with PbBi at MYRRHA relevant conditions, presentation at the MATTER technical meeting, Rome, February 2012.

[55] A. Weisenburger, Corrosion, corrosion barrier development and mechanical properties of steels foreseen as structural materials in liquid lead alloy cooled nuclear systems, presentation at ICONE 17 conference, Bruxelles, July 2009.

[56] R.B. Waterhouse, Fretting Corrosion , I Edition, Oxford, New York, Pergamon Press (1972).

[57] D.A. Hillis, Mechanics of Fretting Fatigue (Solid Mechanics and Its Applications), Kluwer Academic Publishers, Waterloo, Ontario, Canada (2010).

[58] G.W. Stachowiak, A.W. Batchelor, Engineering Tribology, III Edition, Elsevier (2005).

[59] J. Sato, M. Shima, Fretting wear in sea water, Wear, 100 (1986) 227-237.

[60] P.L. Hurricks, The mechanism of fretting – A review, Wear, 15 (1970) 389-409.

[61] M. Antler, Fretting corrosion of gold plated connector contacts, Bell Telephone Laboratories 1979, Columbus, Ohio.

[62] P.L. Hurricks, Some metallurgical factors controlling the adhesive and abrsive wear resistance of steels – A review, Wear, 26 (1973) 285-304.

[63] O. Vingsbo, S. Soderberg, On fretting maps, Wear, 126 (1988) 131-147.

[64] L. Vincent, Y. Berthier, M.C: Dubourg, M.D. Godet, Mechanicms and materials in fretting, Wear, 153 (1992) 135-148.

[65] Ian M. Hutchings, Friction and wear of Engineering materials, CRC Press, 1992.

[66] S. Halliday, The fretting corrosion of mild steel, Proceedings of the Royal Society London, Mathematical and Physical Sciences, A236 (1956) 411-425.

[67] G.H.G. Vaessen, Proceeedings of the Institute of Mechanical Engineering, London, 183 (1968-1969) 125.

[68] N. Ohmae, T. Tsukizoe, The effect of slip amplitude on fretting, Wear, 27 (1974) 281.

[69] S.C. Gordelier, T.C. Chivers, A literature review of palliatives for fretting fatigue, Wear, 56 (1979) 177.

[70] Z.R. Zhou, Nucleation and early growth of tribologically transformed structure (TTS) induced by fretting, Wear, 212 (1997) 50-58.

[71] G.Z. Xu, The effect of the third body on the fretting wear behavior of coatings, Journal of Materials Engineering and Performance, 11 (2002) 288-292.

[72] N. Diomidis, Third body effects on friction and wear during fretting of steel contacts, Tribology International, 44 (2011) 1452-1460.

[73] J.M. Dobromirsky, Third body effects on friction and wear during fretting of steel contacts, *Standardization of Fretting Fatigue Test Methods" and Equipment, ASTM STP* 1159, M. Helmi Attia and R. B. Waterhouse, Eds., American Society for Testing and Materials, Philadelphia, 1992, 60-66.

[74] E.S. Sproles, The mechanism of material removal in fretting, Wear, 49 (1978) 339-352.

[75] B. R. Pearson, P.A. Brook, R.B. Waterhouse, Fretting in aqueous media – particularly of roping steels in seawater, Wear, 106 (1985) 225-260.

[76] G. Loffieial, L'usure dans Ies cables et conduits flexible - une etude de cas en tribologie, Em&b, September 1985, Vol. 3, Ecole Centrale de Lyon, pp. 221-225.

[77] C. Chamont, Y. Honnorat, Y. Berthier, M. Godet, L. Vincent, Wear problems in small displacements encountered in titanium alloy parts in aircraft turbomachines, World Conf. on Titanium, June 6-9, Z988, Cannes. D.H. Jones, A.Y. Nehru, J. Skinner, The impact fretting wear of a nuclear reactor component, Wear, 106 (1985) 139-162.

[78] C. Cattaneo, Sul contatto di due corpi elastici: distrbuzione locale degli sforzo, Rendiconti dell´Accademia Nazionale dei Lincei, Vol 27, Ser. 6, 1938.

[79] R.D. Mindlin, Compliance of elastic bodies in contact, Journal of Applied Mechanics, 16 (1949) 259-268.

[80] K.L. Johnson, Surface interaction between elastically loasde bodies under tangential force, Proceedings of the Royal Society London, Seria A, 230 (1955) 531-548.

[81] J. Sato, M. Shima, T. Sugawara, A fundamental study of fretting damage to glass using an improved apparatus, Wear, 106 (1985) 53-61.

[82] R.C. Bill, Fretting of Nichel-Chromium-Aluminium alloys at temperatures to 816°C, NASA TN D-7570, 1974.

[83] E. Sauger, L. Ponsonnet, J.M. Martin, Study of the tribologically transformed structure created during fretting tests, Tribology International, 33 (2000) 743-750 .

[84] R.B. Waterhouse, Fretting debris and delamination theory of wear, Wear, 29 (1974) 337-344.

[85] S. Jahanmir, N.P. Suh, Mechanism of subsurface void nucleation in delamination wear, Wear, 44 (1977) 17-38.

[86] N.P. Suh, An overview of delamination theory of wear, Wear, 44 (1977) 1-16.

[87] U. Bryggman, S. Soderberg, Contact conditions in fretting, Wear, 110 (1986) 1-17.

[88] J. Warburton, The fretting of mild steel in air, Wear, 131 (1989) 365-386.

[89] M.P. Szolwinski, Mechanics of cracks formation, Wear, 198 (1996) 93-107.

[90] J.R. Fleming, N.P. Suh, Mechanics of crack propagation in delamination wear, Wear, 44 (1977) 39-56.

[91] Y. Berthier, L. Vincent, M.C. Dobourg, Fretting fatigue and fretting wear, Tribology International, 22 (1989) 235-242.

[92] R.B. Waterhouse, Fretting fatigue, Applied Science Publishers Ltd., London, 1981.

[93] R.B. Waterhouse, Fretting fatigue, International Materials Reviews, 37 (1992) 77-95.

[94] K. Endo, H. Goto, Initiation and propagation of fretting fatigue cracks, Wear, 38 (1976) 311-324.

[95] D.A. Hills, D. Nowell, J.J. O'Connor *et al.*, On the mechanics of fretting fatigue, Wear, 125 (1988) 129-146.

[96] K. Sato, H. Fujii *et al.*, Crack propagation behavior in fretting fatigue, Wear, 107 (1986) 245-262.

[97] H.H. Uhlig, Mechanism of Fretting, Journal of Applied Mechanics, 21 (1954) 401-407.

[98] D.E. Taylor, The fretting wear of an austenitic stainless steel in air and in carbon dioxide at elevated temperatures, Wear 56 (1979) 9-18.

[99] J.C. Killeen, A.F. Smith, R.K. Wild, Chromium depletion profiles after preferential removal of chromium from alloys, Corrosion society, 16 (1976) 551-559.

[100] B. Vick, M.J. Furey, A basic theoretical study of the temperature rise in sliding contact with multiple contacts, Tribology International, 34 (2001) 823-829.

[101] J.F. Matlik, High frequency, high-temperature fretting-fatigue experiments, Wear, 261 (2006) 1367-13842.

[102] K. Endo, H. Goto, Effects of environment on fretting fatigue, Wear, 48 (1978) 347-367.

[103] S. Soderberg, U. Bryggman, T Mc Collough, Frequency effects in fretting wear, Wear, 110 (1986) 19-34.

[104] E. Sauger, S. Fouvry, L. Ponsonnet, P. Kapsa, J.M. Martin, L. Vincent, Tribologically transformed structure in fretting, Wear, 245 (2000) 39-52.

[105] M. Verenberg, G. Halperin, I. Etsion, Different aspect of the role of wear debris in fretting wear, Wear, 252 (2002) 902-910.

[106] R. Ramesh, R. Gnanamoorthy, Effect of hardness on fretting behavior of structural steel, Material Design, 28 (2007) 1447-1452.

[107] P. Blanchard, C. Colombie, V. Pellerin, S. Fayeulle, L. Vincent, Material effects in fretting wear: application to iron, titaitum and aluminium alloys, Metal Transaction, 22 (1991) 1535-1544.

[108] I. Sherrington, P. Hayhurst, Simoultaneous observation of the evolution of debris density and friction coefficient in dry sliding steels contacts, Wear, 249 (2001) 182-187.

[109] R.E. Pendlebury, Formation, readhesion and escape of wear particles in fretting and sliding wear in inert and oxidizing environments, Wear, 125 (1988) 3-23.

[110] A. Iwabuchi, The role of oxide particles in the fretting wear of mild steel, Wear, 151 (1991) 301-311.

[111] T.J Benham, E.R. Leheup, J.R. Moon, The effect of debris chemistry on the debris formation and fretting behavior of mild steel in CO_2 containing oxygen and water vapor, Wear, 177 (1994) 195-202.

[112] C. Colombie, Fretting: load carrying capacity of wear debris, Wear, 106 (1984) 194-201.

[113] M. Godet, The third-body approach: a mechanical view of wear, Wear, 100 (1984) 437-452.

[114] J. Vizintin, Three-body contact temperature in fretting conditions, Symposium on Tribology, Lyon, France, 1995.

[115] D.A. Rigney, Wear processes in sliding systems, Wear, 100 (1984) 195-219.

[116] M Sawa, D.A. Rigney, Sliding behavior of dual phase steels in vacuum and in air, Wear, 119 (1987) 369-390.

[117] P Heilmann, J. Don, T.C. Sun, D.A. Rigney, W.A. Glaeser, Sliding wear and transfer, Wear, 91 (1983) 171-190.

[118] J.S. Benjamin, The mechanism of mechanical alloying, Metal Transactions, 5 (1974) 1929-1934.

[119] R. Österlund, Butterflies in fatigued ball-bearings; formation mechanisms and structure, Journal of Metals, 11 (1982) 23-32.

[120] I. Feng, Journal of Applied Mechanics, 21 (1954) 395

[121] I. Feng, Proceedings Institute of Mechanical Engineering, 170, (1956) 1055.

[122] C.P.L. Commissaris, Proceedings Institute of Mechanical Engineering, 181, (1967) 3.

[123] D.G. Kim, Y.Z. Lee, Experimental investigation on sliding and fretting wear of steam generator tube materials, Wear, 250 (2001) 673-680.

[124] O. Jin, Shear force effects on fretting fatigue behavior of Ti-6Al-4V, Metals and Materials Transactions, 35A (2004) 131-138.

[125] J.S. Halliday, Proceeedings of Conference on Lubricants and Wear, Institutes of Mechanical Engineering (1957).

[126] R.R. McMath, Transaction of American Society for Lubricants Engineering, 4 (1961) 199.

[127] Y.W. Park, S.N. Narayanan, K.Y. Lee, Effect of fretting amplitude and frequency on the fretting corrosion behavior of tin plated contacts, Surface & coatings Technology, 201 (2006) 2181-2192.

[128] P. Schaaf, Frequency effect in fretting wear of Co-28Cr-6Mo versus Ti-6Al-4V implant alloys, Journal of Biomedicals Materials research Part B: Applied Biomaterials, 77 (2006) 9-88.

[129] P.L. Hurricks, The fretting wear of mild steel from room temperature to 200°C, Wear, 19 (1972) 2001-229.

[130] P.L. Hurricks, The fretting wear of mild steel from 200° to 500°C, Wear, 30 (1974) 189- 212.

[131] T.F.J. Quinn, Oxidational wear, Wear, 18 (1971) 413-419.

[132] R.B. Waterhouse, Fretting at high temperatures, Tribology International, 301 (1981)203-207.

[133] R.C. Bill, Fretting wear and fretting fatigue – How are they related?, Transaction ASME, Journal of Lubrication technology, 105 (1983) 220-238.

[134] R.C. Bill, Fretting of AISI 9310 stel and selected fretting resistant surface treatments, ASLE Transactions, 21 (1977) 236-242.

[135] N.N. Krasnoshchenkov., High temperature friction, Wear Resistant Materials, Kiev, 1969.

[136] E.F. Bisson, R.L. Johnson, M.A. Swikert, Friciton, wear and surface damage of metals as affected by solid surface films, NACA TN 3444 (1955).

[137] R.B. Waterhouse, A. Iwabuchi *et al.*, High temperature fretting of four titanium alloys, Wear, 106 (1985) 303-313.

[138] M.M. Hamdy, R.B. Waterhouse, The fretting wear of Ti-6Al-4V and aged Inconel 718 at elevated temperatures, Wear, 71 (1981) 237-248.

[139] R.C. Bill, The role of oxidation in the fretting wear process, NASA TM-81570, AVRADCOM-TR-80-C-15 (1980).

[140] J. Sato, M. Shima, T. Sugawara, A. Tahara, Effect of lubricants on fretting wear of steel, Wear, 125 (1988) 83-95.

[141] A. Neyman, The influence of oil properties on the fretting waer of mild steel, Wear, 152 (1992) 171-181.

[142] Y. Qiu, BJ. Roylance, The effect of lubricant additives on fretting wear, Lubrication engineering, 48 (1992) 801-808.

[143] J.R McDowell, Fretting of hardened steels in oil, ASLY transactions, 1 (1958) 287-295.

[144] M. Imai, H. Teramoto, Y. Shimauchi, E. Tonegawa, Effect of oil supply in fretting wear, Wear, 110 (1986) 217-225.

[145] Z.R. Zhou, L. Vincent,, Lubrication in fretting – A review, Wear, 225 (1999) 962-967.

[146] Y. Shen, Fretting wear behaviors of steel wires under friction increasing grease conditions, Tribology International, 44 (2011) 1511-1517.

[147] T. Kayaba, A. Iwabuchi, influence of hardness on fretting wear, Wear, 66 (1981) 27-41.

[148] T. Kayaba , A. Iwabuchi, The fretting wear of 0.45% C steel and austenitic steel from 20 to 650°DC in air, Wear, 74 (1981-1982) 229-245.

[149] P.W. Leech, A.W. Batchelor, G.W. Stachowiak, Laser surface alloying of steel wire with chromium and zirconium, Journal of Materials Science Letters, 11 (1992) 1121-1123.

[150] A.W. Batchelor, G.W. Stachowiak, Control of fretting friction and wear of roping wire by laser surface alloying and physical vapour deposition coatings, Wear, 152 (1992) 127-150.

[151] K.J. Kubiak, T.G. Mathia, Influence of roughness on contact interface in fretting under dry and boundary lubricated sliding regimes, Wear, 267 (2009) 315-321.

[152] K.J. Kubiak, T.G. Mathia, S. Fouvry, Interface roughness effect on friction map under fretting contact conditions, Tribology International, 43 (2010) 1500-1507.

[153] P.L. Hurricks, The effect of temperature on the fretting wear of mild steel, Proceedings of the Institute of Mechanical Engineering. London, 184 (1969-70) 165-175.

[154] M. Dalmiglio, P. Schaaff, U. Holzwarth, The effect of surface treatment on the fretting behavior of Ti-6Al-4V alloy, Journal of Biomedicals Materials research Part B: Applied Biomaterials, (2007) 407-416.

[155] C.H. Hager Jr., J.H. Sanders, S. Sharma, Unlubricated gross slip fretting wear of metallic plasma-sprayed coatings for Ti6Al4V surfaces, Wear, 265 (2008) 439-451.

[156] Z.R. Zhou, K. Nakazawa, M.H. Zhu, N. Maruyama, Ph. Kapsa, L. Vincent *et al.*, Progress in fretting maps, Tribology International, 39 (2006) 1068-1073.

[157] Q.Y. Liu, Z.R. Zhou, Effect of displacement amplitude in oil-lubricated fretting, Wear, 239 (200) 237-243.

[158] M. Helmi Attia, Freting fatigue and wear damage of structural components in nuclear power stations – Fitness for service and life management perspective, Tribology International, 39 (2006) 1294-1304.

[159] A. Naviglio, Fluid-induced vibration in the fuel bundle of a PbBi reactor, Transactions Smirt 16, Washington 2001.

[160] H.K Kim, Y.H. Lee,, Wear depth model for thin tubes supports, Wear, 263 (2007) 532-541.

[161] N.J. Fisher, M.K. Weckwerth, D.A.E. Grandison, B.M. Cotnam, Fretting wear of zirconium alloys, Nuclear Engineering and design, 213 (2002) 79-90.

[162] M. Helmi Attia, On the fretting wear of Zr-alloys, Tribology International, 39 (2006) 1320-1326.

[163] S.H. Jeong, J.H. Lee, I.K. Park, S.J. Song, M.Y. Choi, Wear transitions of tube support materials for a nuclear steam generator through sliding wear test and fretting wear test, Key Engineering Materials, 321-323 (2006) 430-433.

[164] M.K. Lim, S.D. Oh, Y.Z. Lee, Friction and wear of Inconel 690 and Inconel 600 for steam generator tube in room temperature, Nuclear Engineering and Design, 226 (2003) 97-105.

[165] A.H. Lee, A comparative study on fretting wear of steam generator tubes in Korean power plants, Wear, 225 (2003) 1198-1208.

[166] Y.H. Lee, Effect of spring shapes on the variation of loading conditions and wear behavior of the nuclear fuel rod during fretting wear tests, Wear, 263 (2007) 451-457.

[167] M.J. Pettogrew, L.N. Carlucci, C.E. Taylor, N.J. Fisher, Flow-induced vibration and related technologies in nuclear components, Nuclear Engineering and Design, 131 (1991) 81-100.

[168] S. Odahara, Fatigue failure by in-line flow-induced vibration and fatigue life evaluation, JSME International Journal. 48 (2005) 109-117.

[169] Y.S. Shin, M.W. Wambsganss, Flow induced vibration in LMFBR steam generators: a state of the art review, Nuclear Engineering and Design, 40 (1977) 235-284.

[170] K. Fujita, Flow induced vibration and fluid structure interaction in nuclear power plants components, Journal of Wind Engineering and Industrial Aerodynamics, 22 (1990) 405-416.

[171] Ansaldo Nucleare, Divisione di Ansaldo Energia S.p.a., Internal communication XADS 41 TNIX 72.

[172] R. Ramesh, R. Granamoorthy, Development of a fretting wear test rig and preliminary studies for understanding the fretting wear properties of steels, Material and Design, 27 (2006) 141-146.

[173] [174] A. Ramahlo, J.P. Celis, Fretting laboratory tests: analysis of the mechanical response of test rigs, Tribology Letters, Vol. 14, No. 3, April 2003.

[174] H.H.Uhlig, Test equipment for evaluating fretting corrosion, Anti-Corrosion Methods and Materials, 1 (1954) 247-252.

[175] A. Koenen, Ph. Virmoux, R. Gras, J. Blouet, J.M. Dewulf, J.M. De Monicault, A machine for fretting fatigue and fretting wear testing in cryotechnical and normal environment, Wear, 197 (1996) 192-196.

[176] E. Marui, H. Endo, N. Hasegawa, H. Mizuno., Prototype fretting-wear testing machine and some experimental results, Wear, 214 (1998) 221-230.

[177] G. Müller, A. Heinzel, G. Schumacher, A. Weisenburger, Control of oxygen concentration in liquid lead and lead-bismuth, Journal of Nuclear Materials, 321 (2003) 256-262.

[178] M. Del Giacco, A. Weisenburger, P. Spieler, F. Zimmermann, F. Lang, A. Jianu, G. Mueller, Experimental equipment for fretting corrosion simulation in heavy liquid metals for nuclear applications, Wear, 280-281 (2012) 46-53.

[179] H.U. Wider, J. Carlssonn, E. Loewer,, Renewed interest in lead cooled fast reactors, Progress in nuclear energy, 47 (2005) 44-52.

[180] M. Del Giacco, A. Weisenburger, G. Müller, Fretting corrosion in liquid lead of structural steels for lead-cooled nuclear systems. Preliminary study of the influence of temperature and time, Journal of Nuclear Materials, 423 (2012) 79-86.

[181] E. Rabinowicz, Friciton and wear of materials, 2nd Edition, John Wiley & Sons, Inc., 1995.

[182] F. Noli, P. Misaelides, K. Bethge, Aluminum diffusion in Al-implanted AISI 321 stainless steel using accelerator-based characterization techniques, Nuclear Instruments and Methods in Physics Research, B 139 (1998) 322-326.

[183] C. Zhong, L. Liu, Y. Wu, Y. Deng, B. Shen, B. Shu, W. Hu, Diffusion behavior of aluminum in the surface layer of iron processed by shot peening, Materials Letters, 64 (2010) 1407–1409.

[184] V. B. Trindade, U. Krupp, B.Z. Hanjari, S. Yang, H.J. Christ, Effect of alloy grain size on the high-temperature oxidation behavior of the austenitic steel TP 347, Materials Research, 8 (2005) 371-375.

[185] R. Fetzer, A. Weisenburger, A. Jianu, G. Müller, Oxide scale formation of modified FeCrAl coatings exposed to liquid lead, Corrosion Science, 55 (2012) 213-218.

[186] M. Del Giacco, Structural materials for ADS development – experimental investigation of some aspects of LBE corrosion – Master Thesis Material Engineering Politecnico di Milano / ForschungsZentrum Karlsruhe (FZK), April 2009.

[187] H. Steiner, J. Konys, Oxidation and mass transfer in heavy liquid metal loops, Scientific reports FZKA 7212, April 2006.

[188] G. Rosenblatt, A Wilson, Corrosion by liquid metals, Proceedings of the Fall Meeting of the Metallurgical Society of AIME, 1969, Philadelphia, Pennsylvania.

[189] W. Wu, G.H. Nancollas, A New Understanding of the Relationship Between Solubility and Particle Size, Journal of Solution Chemistry, 27 (1998) 6.

[190] A.K. Shchekin, I. Rusanov, Generalization of the Gibbs-Kelvin-Köhler and Ostwald-Freundlich equations for a liquid film on a soluble nanoparticles, Journal of Chemical Physics, 129 (2008) 154116

[191] K. N. Song, S.B. Lee, S.H. Lee, Performance evaluation of new spacer grid shapes for PWRS, Nuclear Engineering and Technology, 39 (2007) 737-746.

Acknowledgements

At first, I would like to acknowledge Prof. Dr. Oliver Kraft and Prof. Dr. John Jelonnek for accepting the role of thesis main referent and co-referent respectively.

The research activity constituting the basis of this PhD thesis was possible only thanks to the laboratories, the scientific support and the technical expertise provided by the Institute for Pulsed Power and Microwave Technology (IHM) of the Karlsruhe Institute for Technology. In particular, I would like to acknowledge contributions from my collaborators

- Dr. Georg Müller for accepting my candidature for the PhD position, for the continuous support and for being always helpful and understanding during the years I have spent at IHM.
- Dr. Alfons Weisenburger for his scientific support, general supervision of my PhD work and for the cross-curricular knowledge he has shared with me since the first moment I joined the research group by him led (GESA-Group).
- Dr. Renate Fetzer for collaborating in the correction of my PhD thesis and for the valuable advices provided in this regard.
- Dr. Adrian Jianu for the scientific and human support.
- The members of the GESA-Group Frank Zimmermann, Fabian Lang, Patrick Spieler, Dr. Annette Heinzel, Robin Beckers for the continuous technical support.
- Mrs. Gabriela Bertsch: without her it would have been impossible for me to get through the German bureaucracy. She has been always helpful for me.

- The mechanical and the electrical workshops of the institute, without which it would have been possible to realize the FRETHME facility components. In particular I want to thank Harald Zimmermann, Kai-Torsten Schaefer and Klaus Baumann.

The financial support for my PhD work was provided by the EU FP7-212175EU funded GETMAT project. In particular I want to thank Dr. Concetta Fazio, not only for project related matters but also for being always a source of encouragement and improvement.

During my permanence in Karlsruhe I have been far from my family; however they were always able to support me and cheer me up even from distance. To my Father and my Mother, to my grandparents, to my sister Mara and my brother in law Alberto, is dedicated a deep thanks.

A special, huge and lovely thanks is dedicated to my girlfriend, Doroteya Kostadinova, able to withstand me and support me every day, in every moment, even in the most stressing ones.

During the time I spent here, my hometown friends always let me feel their friendship. For this reason I hereby want to express my gratitude to them all, but especially to Marco D´Apruzzo, Stefano Amodio, Giuseppe Carbone, Riccardo Baroni e Davide Galbiati.

I want also to thank also the good friends I have met in Karlsruhe with whom I spent unforgettable moments. In particular I want to remember: Valentino Di Marcello, Ernesto Gonzales, Sylvain Morel, Franck De Bruycker, Guillaume Beauchamp, Gael Menard and Giovanni Pastore.